KEY TO *Coastal & Chaparral* Flowering Plants of Southern California

Revised Fourth Edition

Barbara J. Collins
California Lutheran College
Thousand Oaks, California

Robert H. Goodman Jr.
Citrus College
Glendora, California

Kendall Hunt
publishing company

Front cover image:
Purple Sage, *Salvia leucophylla*, photo by Robert H. Goodman, Jr.
Dedication by Lorence G. Collins © Kendall Hunt Publishing Company
Arroyo Willow, Black Willow, Narrow Leaf Willow, Pacific Willow, & Red Willow
illustrations courtesy of Antonio Anfiteatro
Back cover images:
Shooting Star, *Dodecatheon clevelandii*, & Showy Penstemon, *Penstemon spectabilis*, photos by Herb Adams

Kendall Hunt
publishing company

www.kendallhunt.com
Send all inquiries to:
4050 Westmark Drive
Dubuque, IA 52004-1840

ISBN 978-1-5249-4661-6

Published in the United States of America

DEDICATION

The *Key to Coastal and Chaparral Flowering Plants of Southern California, Fourth Edition* is dedicated to Dr. Barbara J. Collins

(1929–2013)

Dr. Barbara J. Collins began teaching at California Lutheran University (then California Lutheran College) in 1963 after earning her Master of Science and doctorate degrees at the University of Illinois. Inspired by the challenge of learning to identify the new plants she encountered in southern California, she soon decided to share her passion with her students by writing her own plant identification key, which later became the *Key to Coastal and Chaparral Flowering Plants of Southern California*, first published in 1972.

Barbara thoroughly enjoyed teaching botany and biology. She taught for 50 years before passing away on April 30, 2013. She received many awards for her outstanding teaching, authored 10 textbooks, and produced an autobiographical book titled *You Lead a Mean Trail: Life Adventures and Fifty Years of Teaching*. Barbara was beloved by both her students and peers and will always be remembered as a pioneer in the botanical realm. Those who get the opportunity to visit her second home can appreciate her work even more by touring the Barbara Collins Arboretum at California Lutheran University.

TABLE OF CONTENTS

ACKNOWLEDGMENTS

I would like to thank Dr. Barbara Collins and Dr. Larry Collins for their marvelous introductory plant identification book. After using the *Key to Coastal and Chaparral Flowering Plants of Southern California* for the last 15 years in my plant identification class at Citrus College, I have a great appreciation for the time, effort, and dedication put into this book. Special thanks to Sonja Dahl and Antonio Anfiteatro for their help with the illustrations. I would also like to thank Jacob Aragon, Dave Baumgartner, Marian Coensgen-Luna, Denise Pogroszewski, Sierra Sutton, and Alison Yu for their editing recommendations and Herb Adams for his plant photograph contributions.

PREFACE

This book has been produced as an introductory level field guide for the coastal sage scrub and chaparral plants of southern California, excluding grasses and sedges. Plants are identified with the use of a dichotomous key which includes descriptions and drawings for more than 500 species. All taxonomic names have been updated to the latest nomenclature in accordance with *The Jepson Manual: Vascular Plants of California*, second edition (Baldwin, Goldman, Keil, Patterson, Rosatti, and Wilken, 2012).

In this updated key, an attempt has been made to make the terminology more accessible and to aid the user in identification through diagrams and explanations. Following the key are sections entitled **INTRODUCTION TO FLOWERING PLANTS** and **GLOSSARY** which includes additional illustrations and definitions for much of the terminology used.

No effort has been made to keep plant families together, although in some sections they are grouped due to the nature of the key. Family, scientific, and common names are given for each species as well as their description, locality, and any comments concerning edibility, toxicology, or contact irritability. All species are indexed with respect to family, scientific and common name. A plant identification checklist is also included for personal reference.

The purpose of this book is to stimulate an interest in plants and to inspire you to learn more about the positive (edibility, carbon sequestering, oxygen producing) and negative (poisonous) interactions which humans and plants may have with one another. You may come away with a fresh and unique perspective about plants that you never had before, and find that a curiosity has been quenched or made even deeper. Whatever your experiences are after using this book, it is our hope that you will have a new appreciation for the coastal and chaparral plants of southern California.

AREA OF COVERAGE

The area encompassed by this plant identification key extends from Santa Barbara County to San Diego County and includes common native and non-native vegetation under 4,000 feet on the ocean side of the mountains and between 3,000 and 5,500 feet on the desert side of the mountains. Vegetation restricted to the deserts and higher mountain slopes is excluded, but some chaparral species will overlap into these regions. Some species will also overlap into areas north of Santa Barbara and south of San Diego.

See map below for the general area of coverage.

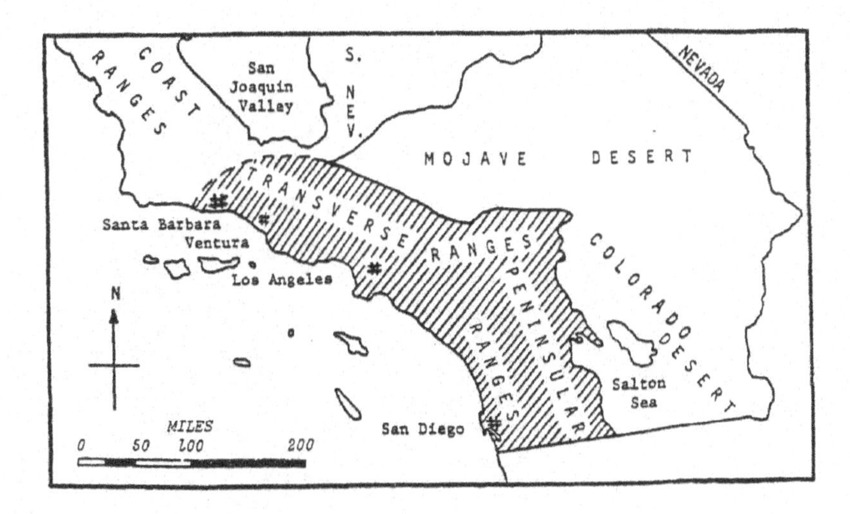

SHADED AREA SHOWS REGION COVERED BY KEY IN SOUTHERN CALIFORNIA

HOW TO USE THE FLOWERING PLANTS KEY

The key contained in this book is dichotomous, meaning two alternatives are always present from which to choose an appropriate plant characteristic. Always be sure to read both choices before making a decision.

To use this key, you should have in hand a representative portion of the plant you wish to identify, including both leaves and flowers, not just flowers or leaves alone. Your memory may not always be accurate and may lead you to the wrong plant in the key. If the plant is a tree or shrub, this should be noted along with the relative size; for example, a tall tree, a medium-sized bushy shrub, or a low matted shrub.

The dichotomous key starts on page one where your choices will lead you to one of six major sections in the plant key (For a more detailed description of the various sections, review **INTRODUCTION TO SECTIONS OF THE KEY**):

SECTION I: TREES AND SHRUBS
SECTION II: HERBACEOUS DICOTS
SECTION III: COMPOSITES—SUNFLOWER FAMILY (ASTERACEAE)
SECTION IV: VINES, PARASITES, AND CLIMBING PLANTS
SECTION V: CACTI
SECTION VI: MONOCOTS

With the plant in hand you are ready to start keying on page one. The first choice is:

1. Plants woody at the base, trees or shrubs

   or

1. Plants herbaceous, not woody at base

If your plant is herbaceous (not woody at the base), then go to number **2** under this number **1** selection. Now choose which of the second alternatives best suits your plant. Here your choice would be either:

2. Flowers in composite heads of ray and/or disk flowers

   or

2. Flowers not in composite heads

Flowers that are composites are those such as daisies, dandelions, thistles, or some weeds with inconspicuous flowers, such as ragweed. Composites do not have flowers like buttercup, wild rose, or snapdragon which have sepals and petals surrounding a center portion composed of pistil and stamens. If your flower is not a composite, go to number **3** under this number **2** selection which reads:

3. Floral parts in 3's and 6's; leaves parallel veined

   or

3. Floral parts not as above

If your flower has five petals and five stamens, you select the second number **3** alternative. If your flower has six petals, six stamens, and long lily-like leaves with parallel veins, you would go to **SECTION VI: MONOCOTS**. If your flower is not a monocot, proceed to number **4**, and decide if your plant is a vine, a climbing plant, or a parasite. If it is not, proceed to number **5** with the final two choices:

5. Cactus; with spines

or

5. Plants not cactus

Most plants that you will pick probably belong to **SECTION II: HERBACEOUS DICOTS**. To this group belong the herbaceous members of the rose family, the pea family, the snapdragons, and all others that are not specifically cacti, vines, monocots, or composites.

Once you have followed the initial key on page one and have been directed by your choices to a particular section, you will use the same kind of keying procedures in that particular section. In working the key, you should keep the following things in mind:

1. Do not attempt to key out cultivated plants from your garden or the ornamental plants around your house. This key is intended to include only the wild vegetation of the region. However, some weeds and garden plants (non-native exotic plants that have escaped cultivation) which grow wild are included.
2. Always look at both choices before making a decision. When you have made your decision, follow the next consecutive number under that choice.
3. When you finally reach the name of the plant, check the description and illustration with your plant. Do they match? Do they not match? If they do not match, assume that you have made an error some place along the way. You may start over again and try another route, or you may backtrack and see if you can find your mistake. Perhaps there was some spot where you could not quite make up your mind or were not sure. **Note:** there is also the possibility that the plant is not a common plant and is not in this key. If so, you will need to consult a more inclusive reference.
4. If you do not understand a term and the illustrations are not sufficient, utilize the **GLOSSARY** in the back of the book.

When you find the plant which you are trying to identify, you will see an illustration, a brief description of the plant, and the name. You will note that two names are given for the plant. The name in boldface under each plant illustration is the common name. The *italicized* name is the scientific name. You may wonder why scientific names are necessary. First, scientific names are universal and can be used in any language. Second, common names are *not* always the same in all parts of the country. The same plant that is called "tansy" in Washington is called "yarrow" in California. "Tansy" in California is another plant in an entirely different plant family. Third, common names are also sometimes misleading. Spanish moss, for example, which is found draping on trees in Georgia and Florida is not from Spain, nor is it a moss. It is a flowering plant that normally grows on other vegetation.

The scientific name always consists of two names: the first is the genus and the second is the species. The genus refers to a group of plants that have similar characteristics, such as the rose, and is always capitalized. The genus of rose is *Rosa*. Following the genus is the species name for a particular type of rose, such as a California rose, and is not capitalized. The full name then for the California rose would be *Rosa californica*. The scientific name is usually italicized, but when italics are not available, the name should be underlined: Rosa californica.

You will also note that one or more names or initials follow the scientific name; for example: *Rosa californica* Cham. & Schltdl. or *Brassica rapa* L. These names or initials are known as the authority and are the name of the person or persons who gave the plant its name. Usually standard abbreviations are used. Cham. & Schltdl., for example, stand for Adelbert von Chamisso and Diederich Franz Leonhard von Schlechtendal; and L. stands for Linnaeus.

Sometimes a change in nomenclature may be made, such as a change to the genus or family name. In such a case, the name of the original author is placed in parentheses, and the name of the person who made the change is placed next. For example, *Sinapis nigra* was the name given to a particular plant by Linnaeus. Wilhelm Daniel Joseph Koch later changed it to the genus: Brassica. Thus, the name is written: *Brassica nigra* (L.) W.D.J. Koch.

As you identify plants in this key, you may find the plants easier to remember if you take some colored pencils and color the black and white illustrations with the appropriate colors for the flower petals and leaves of the particular plant you have identified. A plant identification checklist has also been provided in the back of the book for this purpose.

INTRODUCTION TO SECTIONS OF THE KEY

SECTION I: TREES AND SHRUBS

This section deals with plants that are definitely woody at the base. Although it may be desirable to have flowers for identification, most species can be keyed out on the basis of leaf characteristics. If you are not familiar with leaf characteristics, check them out in the **INTRODUCTION TO THE FLOWERING PLANTS**, before starting to key. If you do not have flowers on your plant and cannot find the plant in this section, try the tree and shrub portion of the Composite section (**Asteraceae**—sunflower family). It is also a good idea to look for the presence of fruits as they may persist for quite a while on the plant and could give a clue to the identification. Most shrubs that do require flowers for identification in this key are not too noticeable during the winter when flowers are not visible.

Oaks, willows, and some other plants have flowers in **catkins**. Although these may not be on the tree, they may be found lying on the ground underneath the tree. Catkins are unisexual spikes and usually dangle from the trees in early spring.

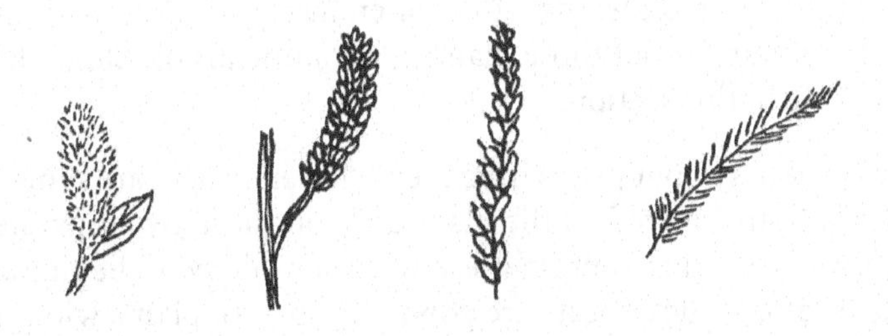

Catkins

The buckwheats (**Polygonaceae**) may be a bit confusing. Although they are often colored, they have no petals. The **sepals** are colored and resemble a corolla. Several of these are usually clustered into a green cup-like structure called an **involucre**. Involucres could resemble a calyx, but a true calyx would only have one flower (corolla) in it and not several as is the case with most buckwheats.

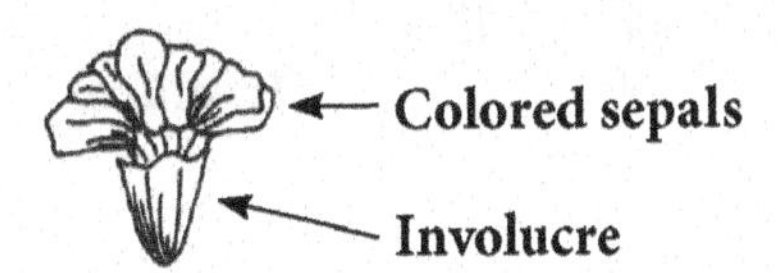

Members of the saltbush family (**Chenopodiaceae**) tolerate considerable alkalinity in the soil. These plants are either fleshy or covered with a scurfy substance which gives them a grayish color. The scurfy substance can be rubbed off with the finger, thus revealing the bright green surface underneath. Flowers of the saltbush family are mostly small, greenish, and globular.

Many plants may be recognized by odor. If you are not familiar with poison oak, however, do not indiscriminately crush a leaf for odor. Poison oak is placed early enough in the key so that this will not happen. If your plant has three leaflets, check for identification before you touch it.

Poison oak

SECTION II: HERBACEOUS DICOTS

This section has the majority of herbaceous plants and includes plants found in water, plants with showy flowers, and many common weeds which often have greenish, inconspicuous flowers. The first division of the key separates out those plants without petals or plants that have tiny flowers. Many common weeds are in this section.

Not all plants in the section without petals have green flowers. There are some plants that are quite showy because the sepals are colored. If the plant does not have two separate sets of floral parts (petals and sepals), the one set that is present is always considered to be the sepals even though it is colored and looks petaloid. Buckwheats are good examples of plants without petals and having colored sepals. The sepals are enclosed in a cup-shaped structure called an involucre which may resemble a calyx.

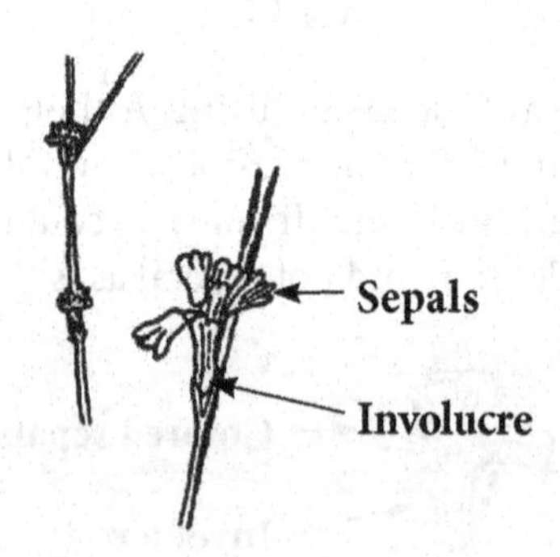

Members of the parsley/carrot/celery family (**Apiaceae**) have all been placed together. These can be recognized by the presence of simple or compound **umbels** and **compound leaves**.

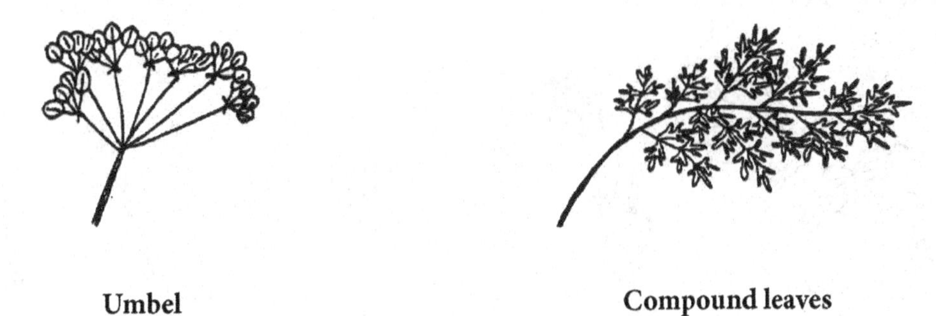

Umbel **Compound leaves**

Members of the pea family (**Fabaceae**) are generally placed together as they all have a characteristic flower which is called **papilionaceous**.

Papilionaceous flower

Members of the mint family (**Lamiaceae**) usually have a distinct odor and a square stem. There are some notable exceptions and these plants are keyed out separately. The hedge nettle, which happens to be a mint, has stinging hairs on the stems, so care should be taken when handling the plant.

One of the real problems for identification can be the showy plants known as Indian paintbrushes (**Orobanchaceae**). The colored parts of the plant are mostly the colored leaves and calyx. The inconspicuous petals are very short, irregular, and are hidden behind the colored leafy bracts. These plants key out in the part for irregular flowers, with particular attention paid to leaf characteristics.

Indian paintbrush

Members of the mustard family (**Brassicaceae**) are also mostly placed together. These may be recognized by the flowers (four petals and six stamens) or by the pod-like fruits (linear or orbicular) that usually occur in terminal racemes.

Mustard family

It must be noted that many plants in this herbaceous section are edible. But, **DO NOT EAT PLANTS**, unless you are sure about their identification and edibility.

SECTION III: COMPOSITES

Composites (**Asteraceae**—sunflower family) are plants whose flowers are arranged in a dense inflorescence called a head. What appears to be one flower is really many flowers compacted together. In the head are two possible kinds of flowers: ray flowers and disk flowers. Ray flowers have one broad petal or corolla which resembles the petals of a single flower. For example, in a daisy what seems like a petal is really a ray flower. The ovary is inferior and is found at the base of the corolla.

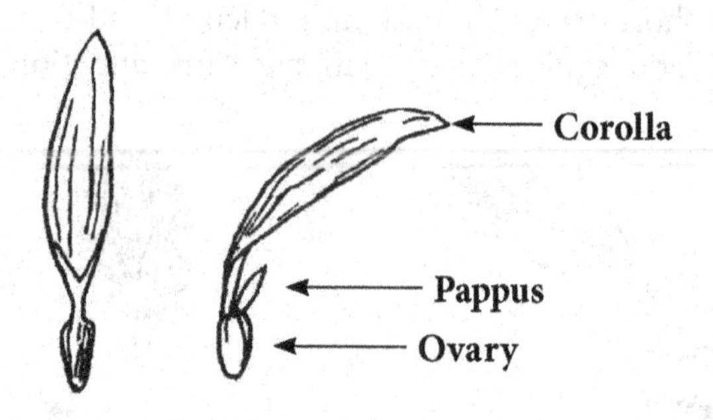

Ray flower

Disk flowers, such as at the center of a daisy, are usually little cylindrical corollas with an inferior ovary. Between the ovary and the corolla may be tiny scales, bristles, or capillary hairs known as the **pappus**. The pappus is best visible when the corollas are gone and the plant is fruiting.

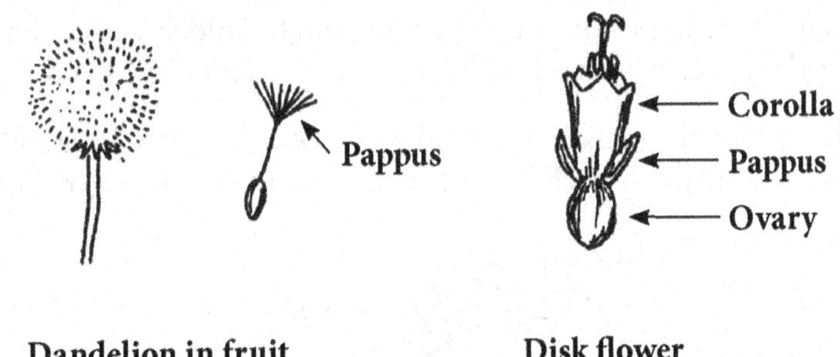

Dandelion in fruit **Disk flower**

On the outside of the composite head are **involucral bracts**. These may be green or papery white, and will be in a single series or have several overlapping rows. Involucral bracts are often a good indicator that a plant is a composite.

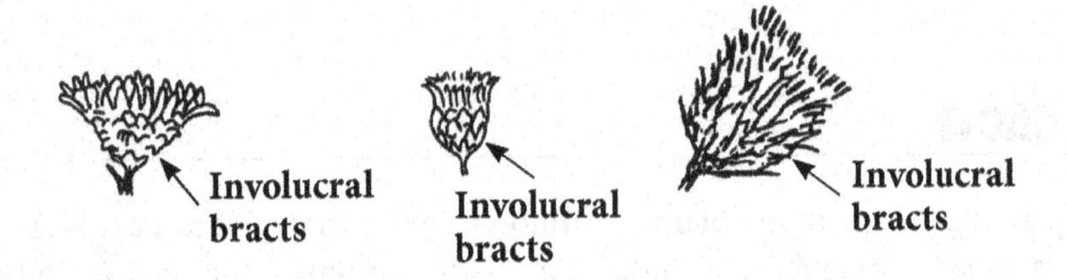

Many composites have both ray and disk flowers. Others have just ray flowers or just disk flowers. Dandelions are examples of plants with just ray flowers. Almost all composites with only ray flowers have milky juice.

Heads with only disk flowers are often greenish and inconspicuous, and may be difficult to recognize, so it is a good idea to scan the illustrations to become familiar with some of the types. For example, ragweed has a cup-shaped involucre enclosing the tiny disk flowers. Some heads with only disk flowers can be colored and quite showy as in *Chaenactis*. The outer disk flowers in *Chaenactis* are enlarged and look almost like rays. Another example of composites with just disk flowers are the thistles, which are easily identified by their spiny bracts and leaves.

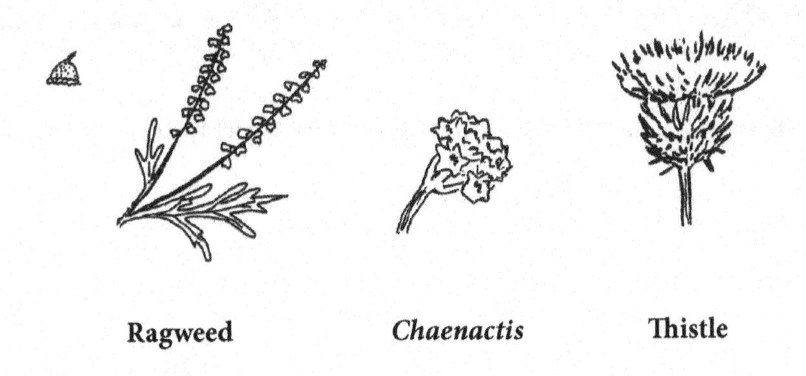

Ragweed ***Chaenactis*** **Thistle**

SECTION IV: VINES, PARASITES, AND CLIMBING PLANTS

Parasitic plants get their nourishment from other plants and may or may not have chlorophyll. Mistletoe is an example of a hemiparasite with chlorophyll. Dodder is devoid of chlorophyll, is bright orange and almost resembles a fungus. Even though dodder lacks chlorophyll, it is recognized as a flowering plant because it has tiny flowers and produces fruits and seeds.

Also included in this section are **vines** and **climbing plants**. These often have spring-like extensions of a leaf rachis or stem known as **tendrils**, which aid the plant in climbing over other vegetation.

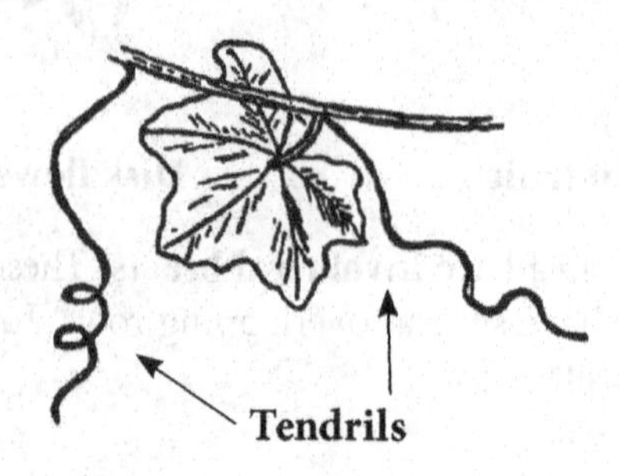

SECTION V: CACTI

Cacti are spiny, succulent, or fleshy plants commonly found in dry desert areas. Several species of cacti occur in the chaparral and are included here. Terminology of the cacti is a bit different.

One characteristic structure of the cacti is the **areole**. These are little spots on the stem from which spines arise. Areoles may be close together or relatively far apart. For example, cholla cacti have areoles quite close together, while prickly pears have areoles farther apart.

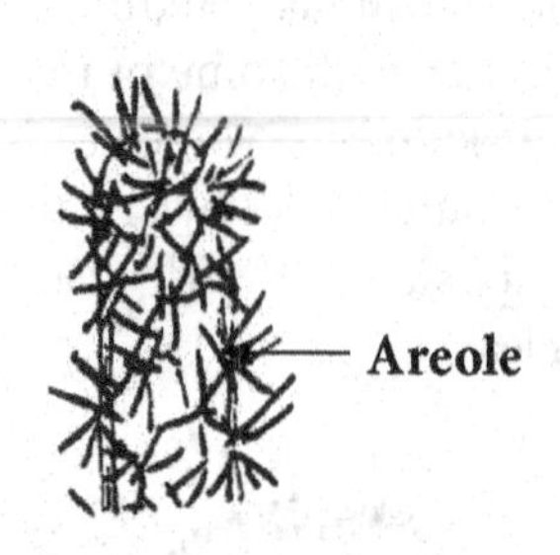

Cholla cacti

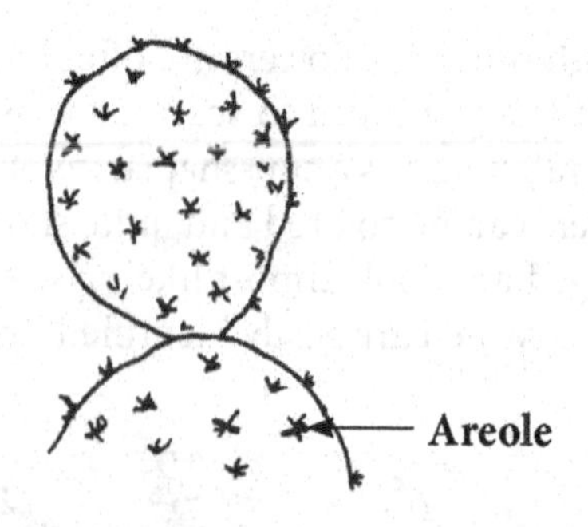

Prickly pear

Most cacti not only have spines in the areoles but also have **glochids**. Glochids are minute barbed bristles which can be just as troublesome as a larger spine. Glochids can be very painful if they get into the skin, and they may be extremely difficult to remove. The best example of glochids is on the beavertail cactus which has no spines at all—just glochids. Prickly pear cacti also have numerous glochids which occur along with the spines in the areoles. Fruits of the prickly pear usually have abundant glochids.

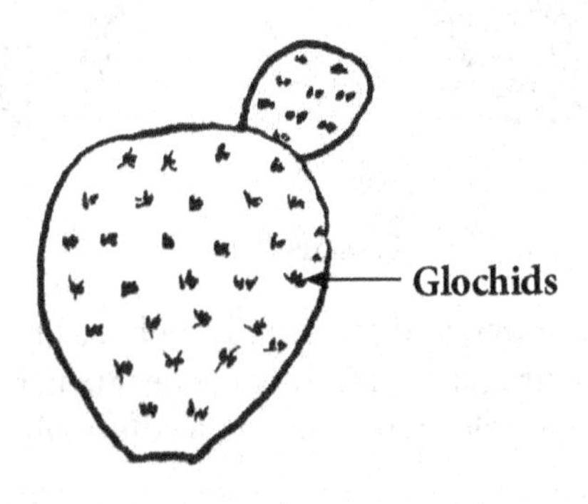

Beavertail cactus

Stems of cacti in the area are either cylindrical as in cholla or flattened as in prickly pear. Some of the larger cacti have a trunk-like base, but this is not common.

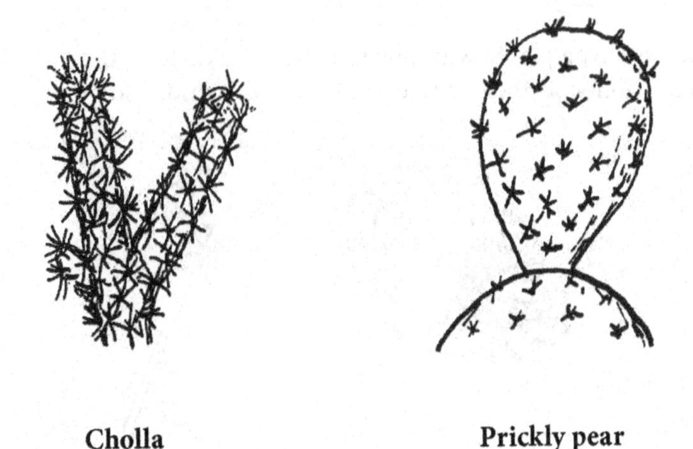

Cholla **Prickly pear**

Spines of the prickly pear have sharp needle points which can be painful if they stab the skin. Spines of cholla, however, are far worse as they are minutely barbed at the tip, like a hundred tiny fishhooks. The spines go into the skin easily but are extremely difficult and painful to remove. It is best to avoid any contact with cholla spines, and because the joints often fall on the ground, care should be taken when you are walking in a cholla patch.

Cactus flowers are usually very showy, having many petals and sepals which are equally colored and indistinguishable. The flowers typically have many stamens and one pistil with an inferior ovary, so the fruits develop beneath the petals and sepals.

Cactus flower

Fruits of the cacti may be fleshy or dry, and are usually globose or pear shaped. Fruits of the prickly pear are fleshy and known for their edibility. Care must be taken, however, to remove all spines and glochids before eating. Pads of the prickly pear are also edible when peeled.

Even though cacti store large amounts of water, it is not possible to cut open a cactus and drink the stored water. All you will find is a fleshy, somewhat juicy tissue with a gritty texture, hardly a substitute for a drink of cool water.

SECTION VI: MONOCOTS

Monocots are those flowering plants with one seed leaf (cotyledon), three or six petals, and long, parallel-veined leaves. Monocots include lilies, onion, iris, orchid, and grasses.

Lily

Onion

Not all monocots have colorful flowers. Flowers of grass are inconspicuous, lacking both petals and sepals. Cat-tail is another example of an inconspicuous flower. However, most other monocots have relatively showy flowers.

Sometimes petals and sepals are both colored and indistinguishable. The petals and sepals are then called perianth parts. Petals and sepals of wild onion are all alike and are called perianth parts, while in the Mariposa lily they are not all alike and are thus called petals and sepals.

KEY TO SECTIONS OF FLOWERING PLANTS

1. Plants woody at the base, trees or shrubs
   2. Flowers in composite heads of ray and/or disk flowers

SECTION III COMPOSITES (Asteraceae—Sunflower Family) **Page 137**

   2. Flowers not in composite heads

SECTION I TREES OR SHRUBS **Page 5**

1. Plants herbaceous, not woody at base
   2. Flowers in composite heads of ray and/or disk flowers (see sunflower illustrations above)

SECTION III COMPOSITES (Asteraceae—Sunflower Family) **Page 137**

   2. Flowers not in composite heads
      3. Floral parts in 3's and 6's; leaves parallel-veined

SECTION VI MONOCOTS **Page 189**

3. Floral parts not as above
  4. Plants vine-like, climbing, or parasitic

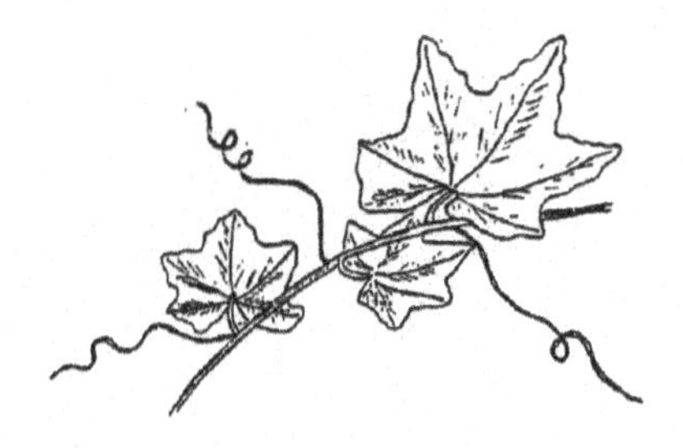

SECTION IV VINES, PARASITES, AND CLIMBING PLANTS **Page 179**

  4. Plants not vines, climbing, or parasitic
    5. Cactus, with spines

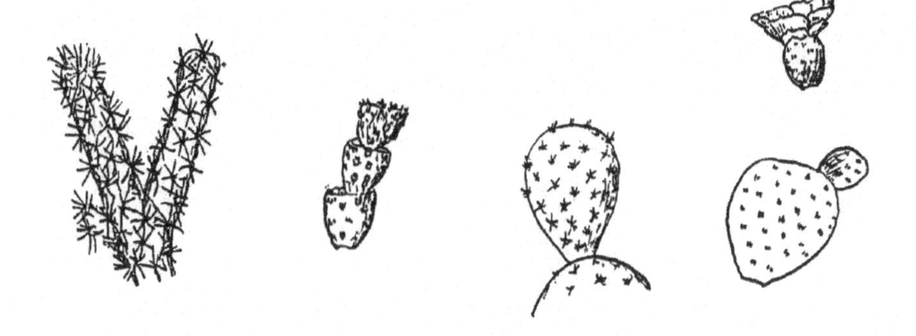

SECTION V CACTI **Page 185**

    5. Plants not cactus

SECTION II HERBACEOUS DICOTS **Page 41**

SECTION I
TREES AND SHRUBS

1. Leaves compound (Leaves simple, page 13)
   2. Leaves pinnate compound (Leaves palmate compound, page 12)
      3. Leaves 2 to 3 times pinnate compound; fernlike

Shrub, 2 to 7 feet high with a strong sweetish odor. Leaves are 2 or 3 times pinnate compound, fernlike in appearance, and glandular pubescent. Flowers are white and occur in a panicle. Petals 5; sepals 5; stamens numerous. Blooms from November to May. Found on dry slopes but not north of San Diego County.

Rosaceae
Chamaebatia australis (Brandegee) Abrams

Southern Mountain Misery

---

3. Leaves once pinnate compound
   4. Leaves 3-foliate (Leaves not 3-foliate, page 10)
      5. Stems prickly
         6. Leaflets toothed. Flowers white

Stems with many slender prickles, twining or scrambling over other plants. Leaves 3-foliate with ovate, dentate leaflets. Flowers in clusters of 2 to 15. Sepals 5; petals 5; white. Stamens many. Inhabits canyons and shaded areas of the chaparral. Blooms from March to July.

Rosaceae
Rubus ursinus Cham. & Schltdl.

California Blackberry

---

6. Leaflets entire. Flowers pink

Evergreen shrub 1½ to 6 feet tall with stiff branches and spines. Leaflets obovate, entire, ¼ to ½ inch long. Flowers showy, purple-pink; solitary. Petals about ¾ inch long. Occasional on chaparral slopes; more frequent north of Santa Monica Mountains. Blooms from May to August.

Fabaceae
Pickeringia montana Nutt.

Chaparral Pea

---

5. Stems not prickly
    6. Leaflets toothed or lobed
        7. Leaves shiny, glabrous. Terminal leaflet with a petiole

Shrub with stems 3 to 6 feet tall. Leaves shining green above; turning dark red in the fall. Flowers small, greenish-white, in axillary panicles, somewhat drooping. Common in shaded, moist areas. Plant secretes a nonvolatile juice which is a **STRONG SKIN IRRITANT** causing **SEVERE BLISTERING AND ITCHING**. Contact is necessary as the juice is non-volatile. The juice may be carried on clothes or vegetation which have contacted the plant or in smoke particles when the plant is burned.

Anacardiaceae
Toxicodendron diversilobum (Torr. & A. Gray) Greene

Western Poison Oak

---

7. Leaves pubescent. Terminal leaflet without petiole

Shrub with stems up to 5 feet tall and pubescent branches. Leaves 3-foliate, pubescent. Leaflets ovate. Flowers yellowish, in clustered spikes. Fruit is reddish, viscid, and hairy. Fruits were used by Native Americans in the preparation of a lemonade-like beverage. Found in canyons. Blooms from March to April.

Anacardiaceae
Rhus aromatica Aiton

Skunk Bush

---

6. Leaflets entire
    7. Flowers regular. Petals 4

Ill-smelling, much branched shrub, 2 to 8 feet tall. Leaves 3-foliate, entire, petiolate. Flowers large, showy, in terminal racemes. Petals 4, bright yellow, about ½ inch long. Sepals 4; stamens 6, long exserted. Fruit a swollen pod, hanging down, ellipitic with pointed tip; 1 to 1¾ inches long. Common shrub in desert or slightly alkaline areas. Also common on bluffs near the coast. Blooms most of the year.

Cleomaceae
Peritoma arborea (Nutt.) H.H. Iltis

Bladderpod

---

7. Flowers irregular, papilionaceous

8. Low bushy shrub, 1 to 4 feet high. Flowers in axillary clusters. Common.

Low bushy shrub 1½ to 4 feet high. Branches slender and straight with a slight soft pubescence. Leaves 3 foliate. Leaflets oblong, ¼ to ⅜ inch long. Flowers yellow, drying reddish; in 1 to 5 flowered umbels. Flowers about ¾ inch long. Fruit a slightly curved pod. A common plant on dry chaparral slopes. Blooms from March to August.

Fabaceae
Acmispon glaber (Vogel) Brouillet **Deerweed**

---

8. Tall shrub, 3 to 10 feet high. Flowers in terminal racemes

9. Branches mostly leafless. Flowers large, ¾ inch long petals

Tall, mostly leafless shrub from 3 to 9 feet tall. Leaves when present with 1 to 3 leaflets. Leaflets oblanceolate, ¼ to ½ inch long. Flowers at ends of leafless, green branches in terminal racemes; bright yellow, fragrant, and showy. Petals ¾ inch long. Pods hairy on margins. Often cultivated and becoming naturalized, mostly north of Santa Cruz. **Spanish Broom**, *Spartium junceum* L., a similar appearing plant is found in the southern mountains. Leaves of Spanish broom are simple rather than compound. Blooms from April to June.

Fabaceae
Cytisus scoparius (L.) Link **Scotch Broom**

---

9. Branches leafy. Flowers smaller, less than ½ inch long

Shrub, 3 to 10 feet tall, with villous branchlets. Leaves 3-foliate with obovate leaflets, ½ to ¾ inch long, pubescent beneath. Flowers in short racemes, 3 to 9 flowered. Petals yellow, less than ½ inch long. Pods hairy all over. Shrub of the coast ranges, from Ventura County north. Blooms from March to May.

Fabaceae
Genista monspessulana (L.) L.A.S. Johnson **French Broom**

---

4. Leaves not 3-foliate. Leaflets 5 to 27

5. Leaflets 5 to 9

6. Stem prickly. Flowers pink to rose. Common

Branched shrub, 3 to 9 feet high with stout curved prickles on stems. Leaves compound with 5 to 7 leaflets. Leaflets oval, serrate, and pubescent beneath. Sepals 5, lanceolate. Petals 5, pink to rose, showy; up to 1 inch long. Stamens numerous. Common in moist areas. Fruits are edible and are high in vitamin C. Flowers from May to August.

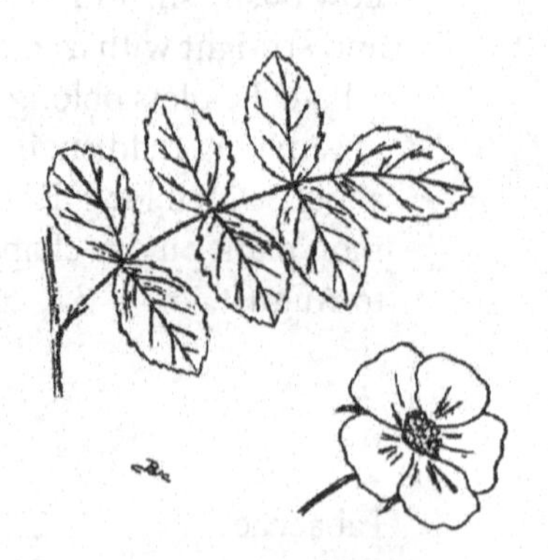

Rosaceae
Rosa californica Cham. & Schltdl.

California Rose

---

6. Stems not prickly

7. Leaflets with spiny teeth. Low shrub 1 to 5 feet high

Evergreen shrub, 1 to 5 feet tall. Leaves pinnate compound with 5 to 9 leaflets that are crowded and overlapping. Leaflets ovate, 1 to 2 inches long with spine-tipped teeth; glossy green above. Flowers bright yellow in drooping racemes. Sepals 6, petal-like. Petals 6. Fruit a berry. Chaparral shrub of canyons and wooded places. Blooms from March to May.

Berberidaceae
Berberis pinnata Lag.

California Barberry

---

7. Leaflets not spiny toothed. Tall shrub or tree more than 5 feet high

8. Fruit a blue berry. Apex of leaflet acuminate

Tree or tall shrub, 6 to 25 feet tall. Leaves pinnate compound with mostly 5 leaflets. Leaflets ovate, finely toothed with acuminate apex. Flowers in a flat-topped spreading inflorescence 1½ to 4 inches across. Flowers small; sepals 5; stamens 5. Corolla 5-lobed. Fruit a blueberry with whitish covering or "bloom." Common tree in canyons and valleys. Blooms April to August.

Adoxaceae
Sambucus nigra L.

Blue Elderberry

---

8. Fruit a samara. Apex of leaflet rounded to acute

Deciduous shrub or small tree, 6 to 20 feet high. Leaves pinnate compound, with 3 to 7 leaflets. Leaflets ovate, toothed, ¾ to 1½ inches long. Flowers appearing before the leaves, forming showy, white clusters. Petals 2, white, inch long. Fruit a samara, ¾ to 1¼ inches long. California ash grows on dry slopes and in canyons of the middle and inner coast ranges. Blooms from March to May.

Oleaceae
Fraxinus dipetala Hook. & Arn.

California Ash

---

5. Leaflets more than 9
6. Leaflets 11 to 15. Tree 10 to 30 feet tall

Deciduous tree, 10 to 30 feet tall. Leaves pinnate compound with 11 to 15 leaflets. Leaflets serrate, lanceolate, 1½ to 3 inches long with an acute apex. Fruit a nut, about 1 inch in diameter, enclosed in a fibrous husk. Common inhabitant of dry hills and valleys of southern California. Blooms from March to May.

Juglandaceae
Juglans californica S. Watson

Southern California Black Walnut

---

6. Leaflets 11 to 27, gland dotted and scented. Shrub 4 to 9 feet high

Deciduous shrub, 4 to 9 feet high with prickle-like glands on young branches and the leaf rachis. Foliage is gland dotted, pubescent and has a heavy scent. Leaves are pinnate compound, 4 to 8 inches long, with 11 to 27 leaflets. Leaflets oblong, entire, ⅜ to 1¼ inches long. Flowers are small and numerous in a long, narrow terminal spike 2 to 8 inches long. Flowers are reduced so only 1 petal is present which is erect, $^3/_{16}$ inch long, and red-purple. Fruit a gland-dotted pubescent pod about ¼ inch long. False indigo grows in dry wooded or brushy slopes and blooms from May to July.

Fabaceae
Amorpha californica Nutt.

California False Indigo

---

2. Leaves palmate compound
  3. Leaves 3-foliate, page 7.
  3. Leaves 4 to 11 foliate
    4. Tree, 20 to 40 feet high. Flowers white to pale rose. Uncommon

Tree, 20 to 40 feet tall. Leaves opposite, palmate compound, with 5 to 7 leaflets. Leaflets lanceolate, toothed, 2 to 6 inches long. Flowers white to pale rose, irregular. Calyx 2-lobed. Petals unequal, clawed, about ½ inch long. Stamens 5 to 7, with orange anthers. Fruit a smooth, pear-shaped capsule. Common in the coast ranges; mostly north of Thousand Oaks. Blooms from May to June.

Sapindaceae
Aesculus californica (Spach) Nutt. **California Buckeye**

---

4. Low shrub, 2 to 6 feet high. Flowers blue-violet to yellow. Common

Low shrub, woody at the base. Leaves palmate compound with 5 to 11 leaflets. Herbage silky pubescent. Leaflets ¼ to ½ inch wide, oblanceolate. Flowers in racemes 4 to 12 inches long. There are several species of this plant that closely resemble one another and may intergrade. **Yellow Bush Lupine,** *Lupinus arboreus,* has slightly larger flowers, averaging ¾ inch long, and occurs along the coast from Ventura County northward. The flowers are usually yellow farther north. **Silver Bush Lupine,** *Lupinus albifrons,* occurs in the north coastal ranges and in the southern interior ranges as far south as San Diego. The flowers average ½ inch long and are whorled on the stem. **Grape Soda Lupine,** *Lupinus excubitus* var. *hallii,* occurs more frequently southward in gravelly, sandy washes. The flowers are blue-violet, averaging ¾ inch long. Leaves and herbage are silvery-white pubescent. On coastal dunes and beaches, **Dune Lupine,** *Lupinus chamissonis,* may be found. Flowers are blue-violet and more or less whorled on the stem. Leaves are silky pubescent. Flowers from March to July. See key on next page:

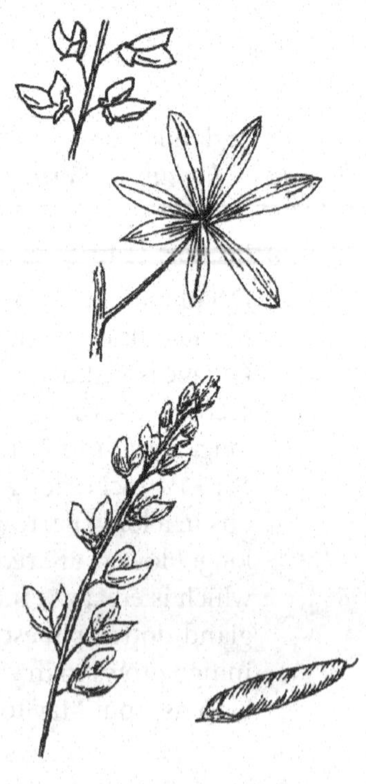

Bush Lupines

---

5. Flowers yellow

Yellow Bush Lupine

Fabaceae
Lupinus arboreus Sims

---

5. Flowers blue-violet
6. Plant of coastal dunes and beaches

Dune Lupine

Fabaceae
Lupinus chamissonis Eschsch.

---

6. Plant of sandy areas away from immediate coast
7. Petioles ¾ to 1½ inches long

Silver Bush Lupine

Fabaceae
Lupinus albifrons Benth.
var. *albifrons*

---

7. Petioles 1½ to 4 inches long

Grape Soda Lupine

Fabaceae
Lupinus excubitus M.E. Jones
var. *hallii* (Abrams) C.P. Sm.

---

1. Leaves simple
2. Leaves opposite or whorled (Leaves alternate, page 19)
3. Leaves whorled
4. Leaves less than 1 inch long. Flowers minute, greenish-white

Shrubby plant 1 to 4 feet high with stiff branches. Leaves linear, acute, ¼ to ¾ inch long, in whorls of 4. Flowers in long panicles; greenish-white, ⅛ inch or less across. Fruit $^{1}/_{16}$ inch in diameter, hirsute with spreading hairs. Shrub of dry slopes or brushy areas. Blooms from April to June.

Rubiaceae
Galium angustifolium A. Gray

Narrowly Leaved Bedstraw

---

4. Leaves more than 1 inch long. Flowers showy, violet-blue
   See *Trichostema lanatum*, page 16.

3. Leaves opposite; not whorled
4. Leaves palmately lobed. Fruit a samara

   Tree, 15 to 100 feet tall. Leaves large, 4 to 10 inches across, deeply 3 to 5 parted. Lobes irregularly toothed. Leaves green above and paler beneath. Flowers small, greenish, in drooping clusters. Fruit a samara with ¾ to 1½ inch wings. Main body of fruit hairy and tawny. Common tree in canyons and along stream banks. Blooms from April to May.

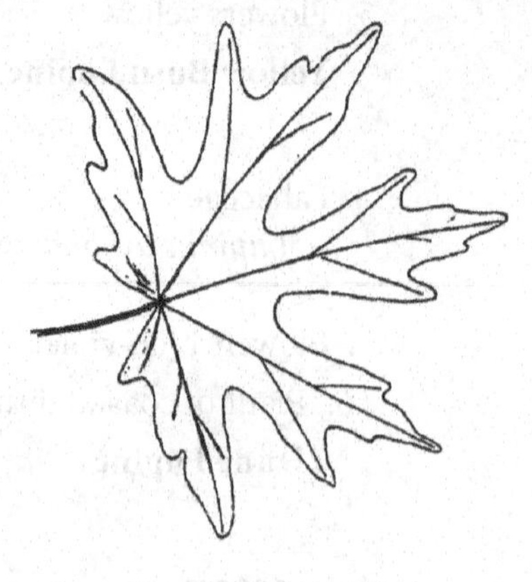
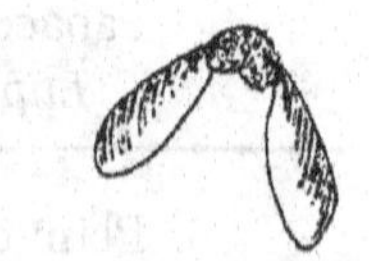

Sapindaceae
Acer macrophyllum Pursh

Big-Leaf Maple

---

4. Leaves not palmately lobed. Fruit not a samara
5. Flowers in a catkin-like inflorescence; without petals

   Shrub 3 to 6 feet tall. Young twigs white tomentose. Leaves ovate, entire, leathery; glabrous and green above; felty and grayish white beneath. Flowers small, apetalous, in catkin-like clusters. Fruit a pubescent berry. Occasional shrub on dry slopes. Blooms from February to April. Also from Ventura County north is **Coast Silktassel**, *Garrya elliptica* Lindl., with more oval-shaped leaves; apex rounded to obtuse rather than acuminate; leaf margin is wavy or undulate.

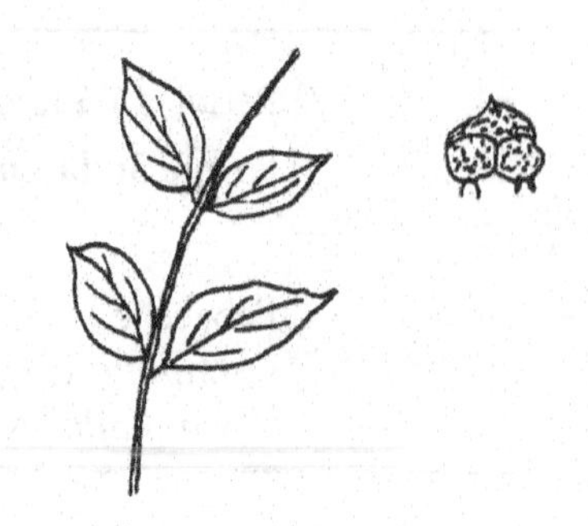

Garryaceae
Garrya veatchii Kellogg

Canyon Silktassle

---

5. Flowers not in a catkin-like inflorescence; petals present
6. Flowers regular; not 2-lipped (Flowers irregular, page 16)
7. Flowers rose or magenta. Leaves heart-shaped

   Low bush, woody at base, with slender, weak stems and viscid pubescence. Leaves ovate, entire, ½ to 1 inch long. Flowers ½ inch long, 5-lobed, each lobe 2-cleft. Stamens 5; exserted. Common bush on dry slopes, blooming from January to June.

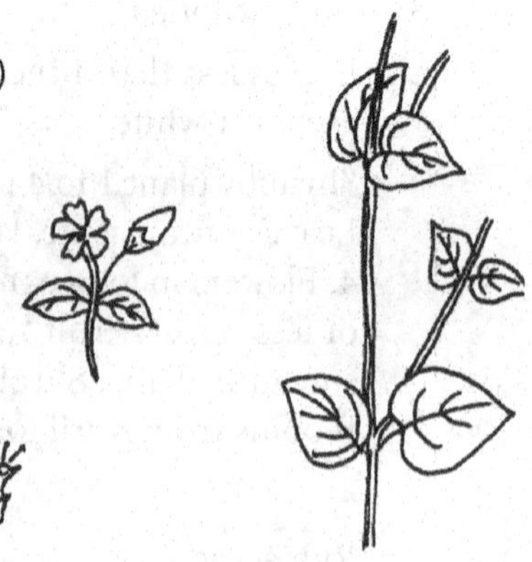

Nyctaginaceae
Mirabilis laevis (Benth.) Curran
var. *crassifolia* (Choisy) Spellenb.

Wishbone Bush

7. Flowers not rose or magenta

8. Low-trailing shrub, 1 to 3 feet tall. Flowers pale pink; bell-shaped

Low trailing shrub 1 to 3 feet tall with tomentose twigs. Leaves opposite, thin, round to oval and entire; densely pubescent beneath. Flowers pink in pairs or small clusters. Ovary inferior. Flowers about ¼ inch long. Fruit a white berry. Occurs in shaded areas and canyons, blooming from April to June. Berries are edible but quite insipid tasting.

Caprifoliaceae
Symphoricarpos mollis Nutt.

Creeping Snowberry

---

8. Erect shrub or small tree, 3 to 18 feet tall. Flowers white

9. Leaves toothed

Shrub 6 to 11 feet tall with grayish tomentose twigs. Leaves revolute, leathery, white tomentose beneath; glabrous above; elliptic, ½ to 1 inch long. Flowers small, white, in lateral umbels. Petals, sepals, and stamens 5. Fruit a sticky capsule with short horns. Common shrub on dry chaparral slopes. Blooms from January to April.

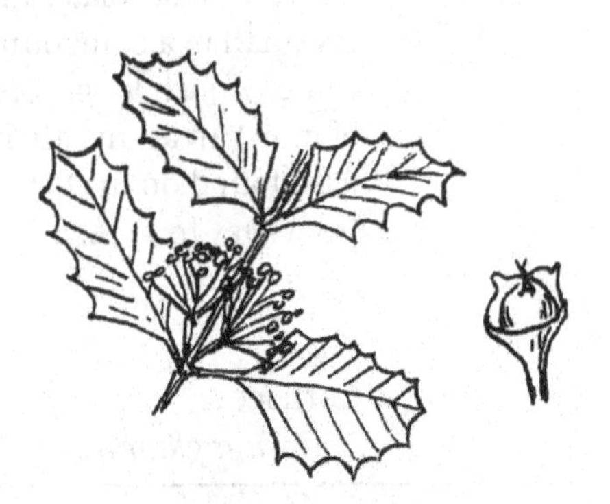

Rhamnaceae
Ceanothus crassifolius Torr.

Hoaryleaf Ceanothus

---

9. Leaves entire

10. Leaves linear, gland dotted; less than ⅛ inch wide

Evergreen shrub 3 to 4 feet tall. Leaves opposite, linear to oblong, and gland dotted; ½ to 1 inch long; entire; and glabrous. The white flowers occur in axillary clusters of 1 to 3 flowers. Petals 4; ¼ inch long. Stamens 8. Calyx 4-lobed. Blooms from November to May and may be found in Orange County south to San Diego.

Rutaceae
Cneoridium dumosum (Torr. & A. Gray) Baill.

Bushrue

---

10. Leaves not linear, gland dotted; leaves more than ⅛ inch wide

11. Leaves spatulate, ¼ to 1 inch long. Common

Rigid shrub 3 to 10 feet tall, commonly found on chaparral slopes. Leaves spatulate, entire, glabrous above and fine pubescent beneath. Leaves firm, averaging ¼ to ½ inch long. Flowers small, white, with 5 petals, sepals, and stamens; occurring in lateral umbels. Fruit a capsule with short, erect horns. A dominant shrub of the chaparral. Blooms from February to May.

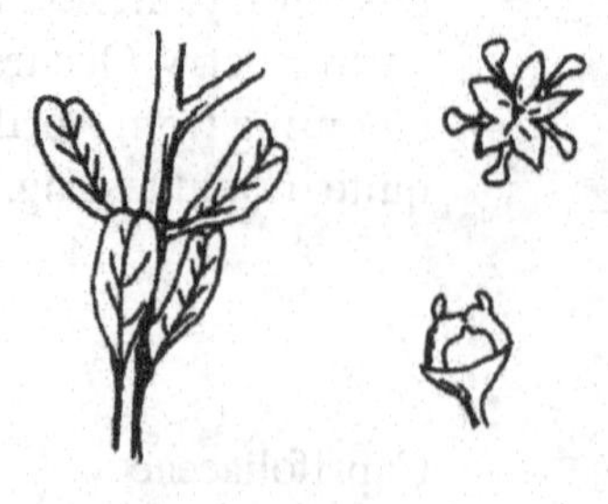

Rhamnaceae
Ceanothus cuneatus Nutt.

Buckbrush

---

11. Leaves lanceolate, 1 to 2 inches long

Tall shrub or small tree 4 to 18 feet tall. Leaves glabrous, lanceolate, 1 to 2 inches long. Flowers small in a compound inflorescence. Petals 4; white, ¼ inch long. Sepals and stamens 4. Shrub of moist areas, mostly north of Ventura County. Also found on Santa Catalina Island. Blooms from May to June.

Cornaceae
Cornus glabrata Benth.

Brown Dogwood

---

6. Flowers irregular; 2-lipped

7. Plants strong odored with a square stem

8. Flowers in distinct axillary whorls

Mint Family

9. Leaves linear; fascicled and often appearing whorled. Flowers covered with a violet and blue wool

Low shrub 1½ to 5 feet tall. Leaves linear-lanceolate with axillary fascicles; revolute, green above and lighter beneath. Flowers in axillary whorls, forming a spike 6 or more inches long. Upper flowering stem and calyx covered with lavender wool. Flowers blue, 2-lipped, covered with blue wool. Stamens 4; long exserted and arched. Common shrub of dry slopes. Blooming from May to August.

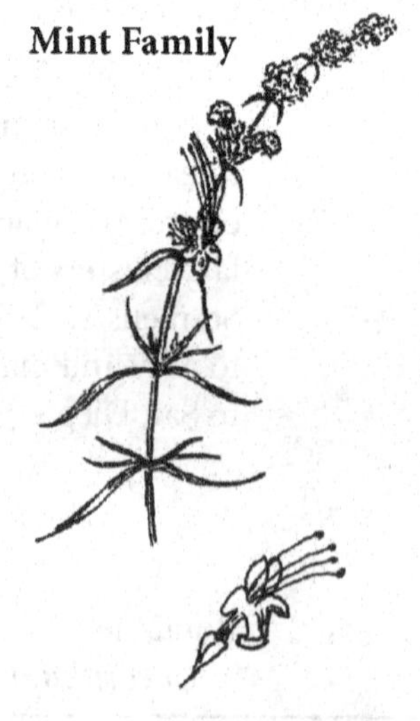

Lamiaceae
Trichostema lanatum Benth.

Woolly Blue Curls

---

9. Leaves not linear or fascicled. Flowers not covered with violet and blue wool

10. Leaves green. Flowers pale blue to white

Shrub 3 to 6 feet tall. Leaves elliptical, green above, slightly tomentose and lighter beneath. Flowers in compact whorls, averaging ¾ to 1½ inch in diameter. Flowers 2-lipped, pale blue to white, ½ inch long. Very common shrub of dry slopes with a strong odor when leaves are crushed. Blooms from April to June.

Lamiaceae
Salvia mellifera Greene

Black Sage

---

10. Leaves and upper stem grayish white. Flowers pale lavender

Shrub 3 to 5 feet tall. Herbage gray tomentose. Leaves oblong-lanceolate, crenulate, with heart-shaped base and prominent veins. Leaves ¾ to 2 inches long. Flowers in compact whorls averaging 1 to 1½ inches across and ¾ inch high, subtended by gray, ovate leaflike bracts. Corolla lavender, ¾ inch long. Common shrub of dry chaparral slopes, blooming from May to July.

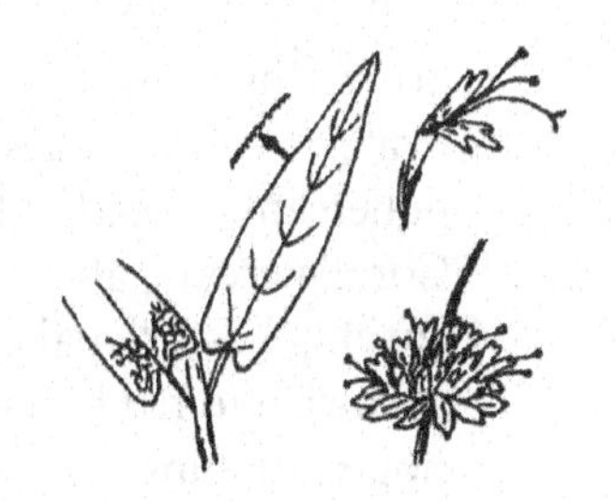

Lamiaceae
Salvia leucophylla Greene

Purple Sage

---

8. Flowers not in axillary whorls

9. Leaves white. Flowers less than ⅞ inch long; in an open panicle on long, willow-like stems

Shrub 3 to 6 feet tall, with long erect branches. Leaves crowded at the base of the present year's growth; opposite, lanceolate. Flowers in compound inflorescences. Petals white to pale lavender, mostly ½ to ¾ inch long with stamens well exserted. Shrub of dry areas with a strong odor. Blooms from April to July.

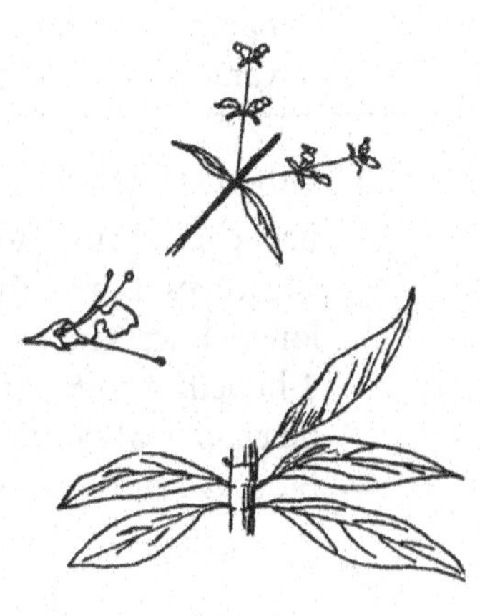

Lamiaceae
Salvia apiana Jeps.

White Sage

---

9. Leaves gray-green. Flowers more than ⅞ inch long; solitary in axils of reduced leaves

Aromatic shrub 1 to 4 feet high. Leaves are ovate-oblong; rather prominently veined and 1 to 5 inches long. Leaf surface is pubescent above and white woolly beneath. The 2-lipped flowers are white with purple veins and markings and are about 1 inch long. Flowers are solitary in axils of reduced upper leaves. Found on open slopes from Ventura County north to Lake County. Flowers April to June. Less common is **Fragrant Pitcher Sage**, *Lepechinia fragrans* (Green) Epling, with deltoid leaves and purplish flowers, found in canyons in Los Angeles County and the adjacent islands.

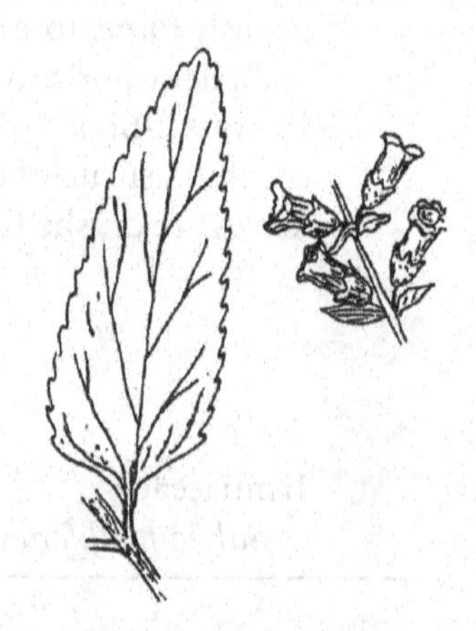

Lamiaceae
Lepechinia calycina (Benth.) Munz

White Pitcher Sage

---

7. Plants without strong mint odor and square stem
8. Flowers yellow- or cream-colored
9. Flowers cream colored. Leaves revolute

Straggling evergreen shrub 3 to 8 feet tall. Leaves ovate, ⅜ to 1¼ inches long, revolute, entire, whitish pubescent beneath. Flowers in short, terminal spikes. Corolla cream, about ¼ inch long, 2-lipped, with lobes curving back. Sepals 5; petals 5; stamens 5. Fruit a yellowish to red berry, ¼ inch across. Shrub of dry slopes. Blooms from April to June. Flowers are often quite fragrant. Berries of this species are edible and may be eaten raw or dried for future use.

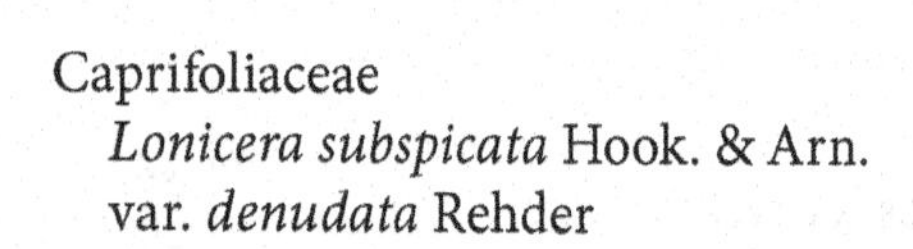

Caprifoliaceae
Lonicera subspicata Hook. & Arn.
var. *denudata* Rehder

Southern Honeysuckle

---

9. Flowers yellow. Leaves not revolute

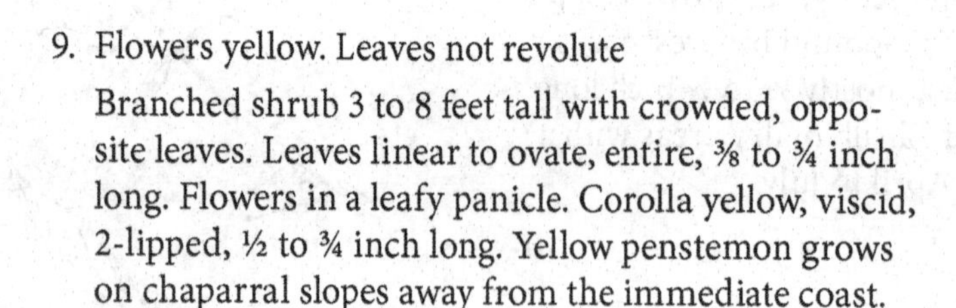

Branched shrub 3 to 8 feet tall with crowded, opposite leaves. Leaves linear to ovate, entire, ⅜ to ¾ inch long. Flowers in a leafy panicle. Corolla yellow, viscid, 2-lipped, ½ to ¾ inch long. Yellow penstemon grows on chaparral slopes away from the immediate coast. Blooms from April to May.

Plantaginaceae
Keckiella antirrhinoides (Benth.) Straw

Yellow Penstemon

---

8. Flowers buff, salmon, orange, or red

9. Leaves lanceolate. Herbage +/– sticky. Common

Shrub 1 to 4 feet high with –/+ sticky pubescent herbage. Leaves lanceolate, 1 to 3 inches long, usually entire and sessile, occasionally with smaller leaves fascicled in the axils. Flowers solitary in the axils, on pedicels ¼ inch long. Calyx ¼ inch long, ribbed and irregularly toothed. Corolla variable: buff, salmon, orange, or red colored; 2 inches long. Very common shrub on rock slopes. Blooms from March to July.

Phrymaceae
Mimulus aurantiacus Curtis

Sticky/Bush Monkeyflower

---

9. Leaves heart-shaped. Herbage glandular only in the inflorescence

Straggling or climbing shrub with stems 3 to 9 feet long. Leaves ovate to cordate or heart-shaped, dentate, shiny. Flowers axillary, drooping, 2-lipped, with wide spreading lower lip. Petals orange-red, 1½ to 2 inches long. Common plant in dry canyons. Blooms from May to July.

Plantaginaceae
Keckiella cordifolia (Benth.) Straw

Heart-Leaved Bush Penstemon

---

2. Leaves alternate

3. Flowers in catkins (Flowers not in catkins, page 24)

4. Fruit an acorn

5. Leaves deeply lobed; deciduous

Large spreading tree 35 to 100 feet tall with light-colored bark. Leaves 2 to 4 inches long, deeply lobed into 3 to 5 pairs of lobes. A common tree in valleys behind the coast ranges. This is the tree that Thousand Oaks was named after.

Fagaceae
Quercus lobata Née

Valley Oak

---

5. Leaves not lobed; evergreen and leathery

6. Shrub 3 to 9 feet tall. Leaves small, less than 1 inch long

Shrub with stiff twigs and evergreen leathery leaves. Leaves dentate with sharp spinose teeth. Leaves ⅝ to 1 inch long, shining above. Common shrub on dry slopes. Blooms from March to May. **Nuttall's Scrub Oak**, *Quercus dumosa* Nutt., found along coast and Santa Catalina Island.

Fagaceae
Quercus berberidifolia Liebm.

Scrub Oak

---

6. Tree 15 to 60 feet tall. Leaves more than 1 inch long

7. Leaves convex on upper surface. Coast ranges

Broad-topped evergreen tree. Leaves convex on upper surface, oval to elliptic, usually sharp or spiny toothed; mostly 1 to 2 inches long. Very common tree of the coast ranges. Blooms from March to April.

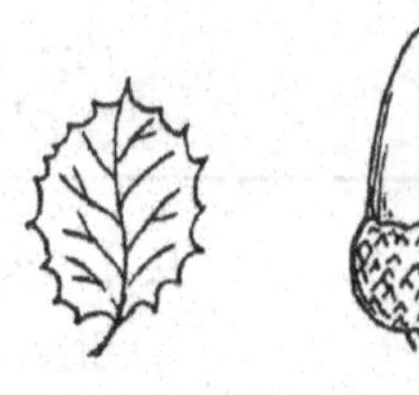

Fagaceae
Quercus agrifolia Née

Coast Live Oak

---

7. Leaves flat on upper surface. Interior

8. Acorn cup densely woolly. Leaves gray or woolly beneath

Evergreen tree, 18 to 60 feet high, with light-colored bark and tomentose young twigs. Leaves, ¾ to 2 inches long; oblong, entire to spinose-toothed, pale green above and gray or yellow tomentose beneath. Fruit an oblong acorn about 1 inch long. Canyon oak is common on slightly moist slopes and in canyons. Catkins in April and May.

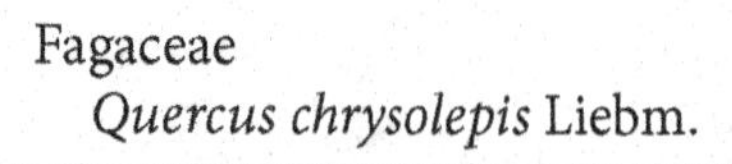

Fagaceae
Quercus chrysolepis Liebm.

Canyon Live Oak

---

8. Acorn cup not densely woolly. Leaves not woolly beneath

9. Leaves dull gray-green, entire to sinuate-toothed

Evergreen tree, 15 to 60 feet high with light-colored bark and tomentose young twigs. Leaves dull gray-green on both surfaces, oblong to obovate, entire to wavy toothed, ¾ to 2½ inches long. Acorn cup enclosing nearly half the nut; pubescent within, with tomentose scales. Acorn ⅝ to 1 inch long.

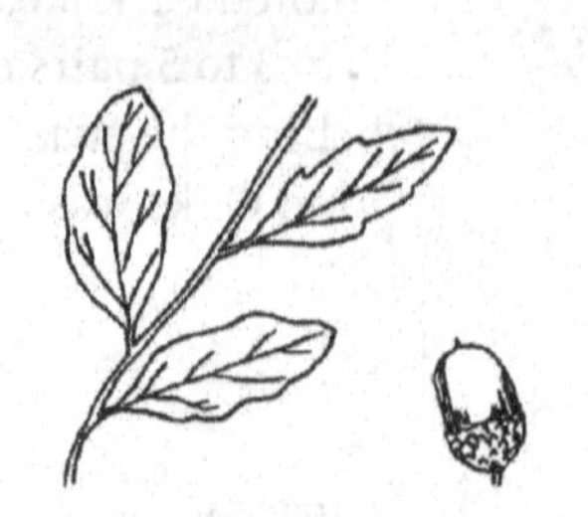

Fagaceae
Quercus engelmannii Greene

Engelmann Oak

---

9. Leaves green, shiny, mostly spiny toothed

Evergreen tree, 30 to 60 feet high, with dark bark which becomes ridged with age. Leaves leathery, oblong to lanceolate, ¾ to 1½ inches long, entire or spiny toothed, pale and shiny above and beneath. Male catkins up to 2¼ inches long. Acorn glabrous, ¾ to 1½ inches long, slender. Acorn cup ⅜ inch deep. Interior live oak is common on slopes and in valleys below 5,000 feet. Flowers from March to May.

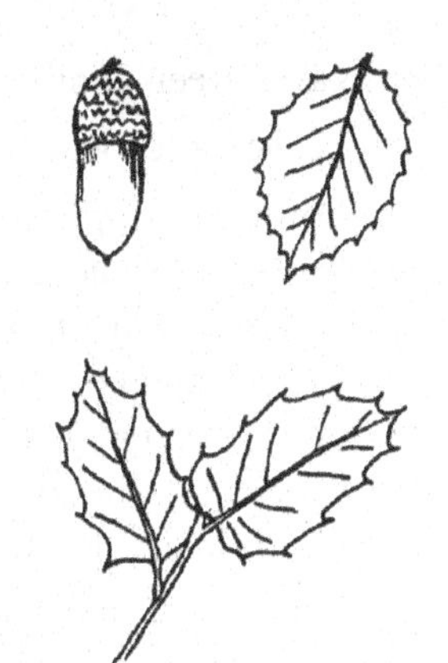

Fagaceae
Quercus wislizeni A. DC.

Interior Live Oak

Willow Family

4. Fruit not an acorn
5. Leaves linear to lanceolate (Leaves broader, page 23)
6. Fruit a capsule. Seeds with a cottony tuft of hairs
7. Leaves linear, 10 times longer than wide

Shrub or small tree, up to 17 feet high, with tomentose young twigs. Leaves linear to linear-lanceolate, tapered at both ends, 10 times longer than wide. Leaf blades entire or toothed, pubescent on both sides. Flowers in catkins, appearing after the leaves. Catkin scales yellow, villous. Capsules villous, ¼ inch long. Narrow-leaved willow is common along ditches, particularly irrigation ditches.

Salicaceae
Salix exigua Nutt.

Narrow-Leaved Willow

---

7. Leaves not as above
8. Leaves same color of green on both upper and lower surfaces

Tree, up to 90 feet high, with yellow to yellow-green, glabrous twigs turning red-brown with age. Leaves finely serrated, alternate, lanceolate, light green above and beneath; to 5 inches long and 1 inch wide. Serrate leaflike stipules maybe present around axillary bud. Flowers in catkins 3¼ inches long. Catkin scales tan. Capsule glabrous. Black willows are common along streams throughout the area. Flowers from March to May.

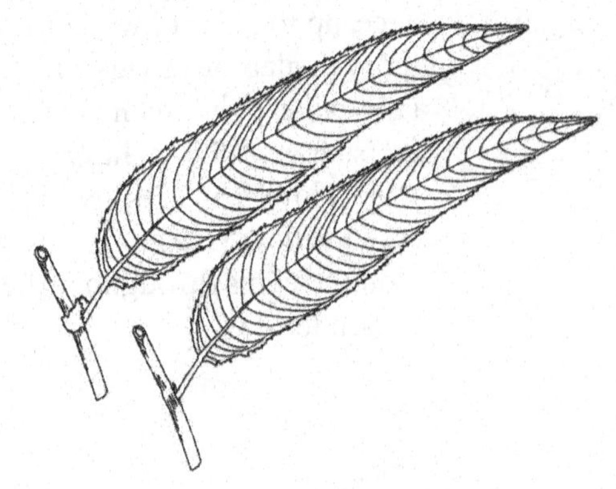

Salicaceae
Salix gooddingii C.R. Ball

Goodding's Black Willow

---

8. Leaves green above, lighter color below

9. Leaves lanceolate to oblanceolate, with leaf tip equal or wider than leaf base

Low spreading tree up to 30 feet tall with somewhat drooping branches. Leaves oblanceolate, green above and lighter blue/green below and 2½ to 4 inches long; leaf tip equal or wider than leaf base. Stipules maybe present on older stems. Common tree along stream banks and dry stream beds. Catkins appear in early spring before the leaves. Blooms February to April.

Salicaceae
Salix lasiolepis Benth.

Arroyo Willow

---

9. Leaves lanceolate, with leaf tip smaller than leaf base

10. Leaf axillary buds rounded at tip; leaf petioles glandular, ¼ to ⅝ inch long. Stipules present and obvious

Tree, up to 36 feet high, with rough brown bark and yellow-gray or red-brown, glabrous twigs. Leaves alternate, lanceolate, finely toothed with acuminate apex, up to 7 inches long, and to 1½ inches wide. Leaf-like glandular stipules present around axillary bud. Male catkins ¾ to 2¼ inches long and female catkins 1¼ to 4 inches long. Catkin scales yellow. Capsule glabrous. Pacific willows are found along stream banks and are not as common as the other willows. Flowers from March to May.

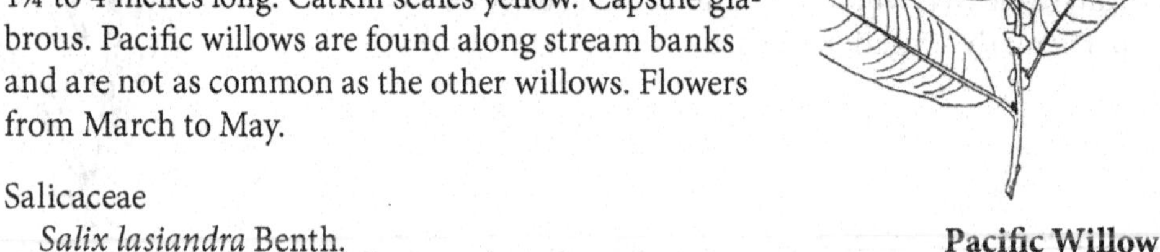

Salicaceae
Salix lasiandra Benth.
var. *lasiandra*

Pacific Willow

---

10. Leaf axillary buds sharp-pointed; leaf petioles smooth, ⅛ to ⅜ inch long. Stipules variable and less obvious

Tree, up to 50 feet high, with yellow to red-brown, hairy to glabrous twigs with age. Leaves alternate, lanceolate, light green above and paler beneath, finely crenate; up to 7.5 inches long and ⅝ to 1¼ inches wide. Flowers in catkins 1 to 4 inches long. Catkin scales yellow. Capsule glabrous. Red willow grows along streams throughout the area. Flowers from March to May.

Salicaceae
Salix laevigata Bebb

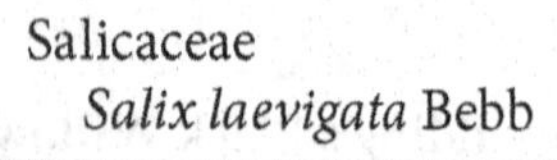

Red Willow

---

6. Fruit a brownish purple nut. Leaves aromatic

Evergreen shrub or small tree 10 to 30 feet high. Leaves simple, alternate, oblanceolate, mostly 2 to 4 inches long and ½ to ¾ inch wide. Leaves dark green and glossy, entire or remotely toothed. Flowers in catkins. The male catkins are in the leaf axils below the female catkins. Fruit a globose, brownish purple nut covered with a whitish wax. Wax myrtle grows on moist slopes and in canyons from the Santa Monica Mountains north. Flowers from March to April.

Myricaceae
Morella californica (Cham. & Schltdl.) Wilbur

Wax Myrtle

---

5. Leaves broader; ovate or deltoid

6. Female fruits forming small cones

Tree 30 to 100 feet tall, with smooth, gray bark. Leaves oblong, ovate, serrate; dark green above, lighter beneath. Leaves 1 to 4¼ inches long. Flowers in catkins. Female catkins short, the scales becoming woody at maturity and resembling tiny cones that are ½ to ¾ inch long. A common tree along streams in canyons and in the mountains. Blooms January to April.

Betulaceae
Alnus rhombifolia Nutt.

White Alder

---

6. Female fruits not forming small cones

7. Leaves deltoid; bright green on both surfaces

Tree 35 to 90 feet tall. Leaves deciduous, deltoid, 1¼ to 2¾ inches long; coarsely serrate. Tree commonly found in moist areas along dry stream beds. Blooms from March to April.

Salicaceae
Populus fremontii S. Watson

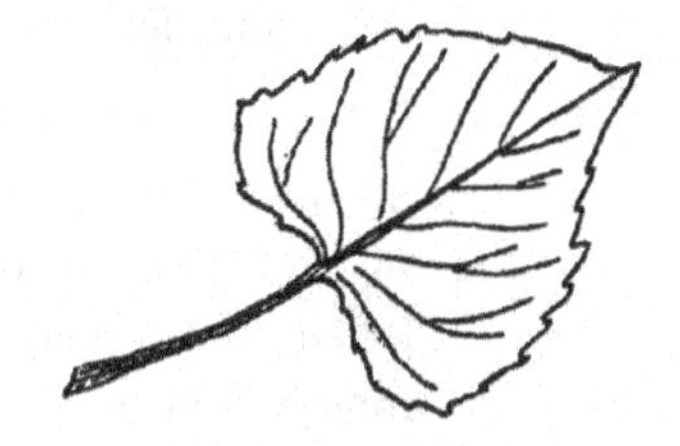

Fremont Cottonwood

---

7. Leaves ovate, dark green above paler beneath

Tree, 100 to 180 feet tall. Leaves serrate with acuminate apex; 1¼ to 2¾ inches long. Common along streams. Blooms from February to April.

Salicaceae
Populus trichocarpa Hook.

Black Cottonwood

---

3. Flowers not in catkins

4. Leaves palmately lobed and veined (Not palmate, page 26)

5. Leaves large, more than 4 inches across

6. Leaves glabrous. Shrub 3 to 9 feet tall

Glabrous shrub with large palmately lobed leaves. Leaves 5 to 11 lobed, the lobes serrate. Leaves 4 to 16 inches across. Flowers, unisexual, without petals. Stamens numerous. Fruit a spiny capsule with smooth mottled seeds. Seeds are **POISONOUS**. An escape from cultivation. Blooms most of the year.

Euphorbiaceae
Ricinis communis L.

Castor Bean

---

6. Leaves tomentose beneath. Tree 30 to 75 feet tall

Tall tree with mottled gray and brown bark. Bark is smooth and peels or exfoliates, producing a camouflaged appearance. Young twigs tomentose. Leaves large, deeply 5-lobed, tomentose beneath; 6 to 10 inches across. Flowers unisexual in globose heads forming little spiny balls in fruit. Balls 1 inch across hanging from trees in strings of 3 or 4. Common in moist areas along stream beds in canyons. Blooms February to April.

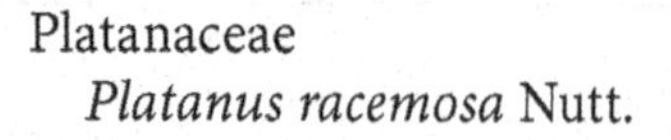

Platanaceae
Platanus racemosa Nutt.

Western Sycamore

---

5. Leaves smaller, less than 4 inches across

6. Plant with spines. Flowers red, drooping

Shrub 3 to 6 feet tall. Branches with 3 spines at the nodes. Spines ½ to ¾ inch long. Leaves palmately lobed, dark green and shiny above, lighter beneath, mostly ½ to 1 inch long. Flowers red, drooping, in groups of 1 to 4. Ovary inferior, glandular, bristly. Sepals and petals 4; ⅜ inch long. Stamens 4, much exserted, about 1¼ or more inches long. Very common plant in shaded canyons. Blooms from January to May. Berries are edible and can be used in jellies and pies.

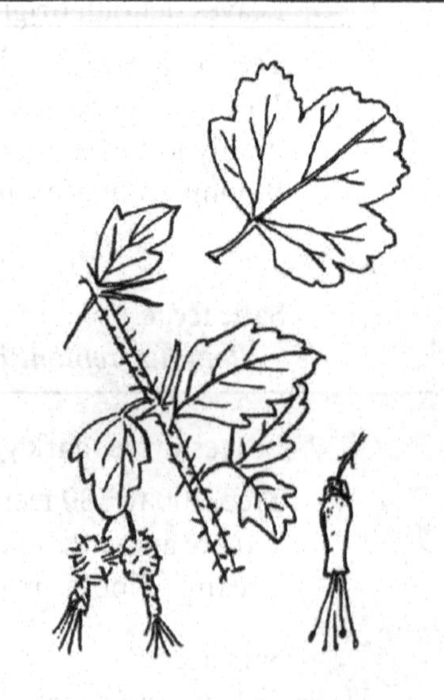

Grossulariaceae
Ribes speciosum Pursh

Fuchsia-Flowered Gooseberry

---

6. Plants without spines

7. Flowers yellow

Shrub 3 to 6 feet high. Leaves glabrous, palmately 3-lobed. Flowers yellow, in racemes. Corolla salverform with a slender tube ¼ to ⅜ inch long. Ovary inferior. Fruit a red or black berry. Chaparral shrub, but not common. Blooms from April to May. Berries are edible and are good for jellies and pie.

Grossulariaceae
Ribes aureum Pursh

Golden Currant

---

7. Flowers not yellow

8. Petals separate. Stamens numerous, fused into a tube around the style

9. Branches woolly. Leaves densely white, tomentose beneath

Shrub 3 to 15 feet tall, tomentose with stellate hairs. Leaves palmately 3 lobed to 5 lobed; gray-green tomentose. Flowers pale pink or lavender in axillary clusters, forming spikes. Petals 5. Stamens numerous, fused into a tube around the style. Common shrub on dry slopes, blooming in the late spring and summer, April to July.

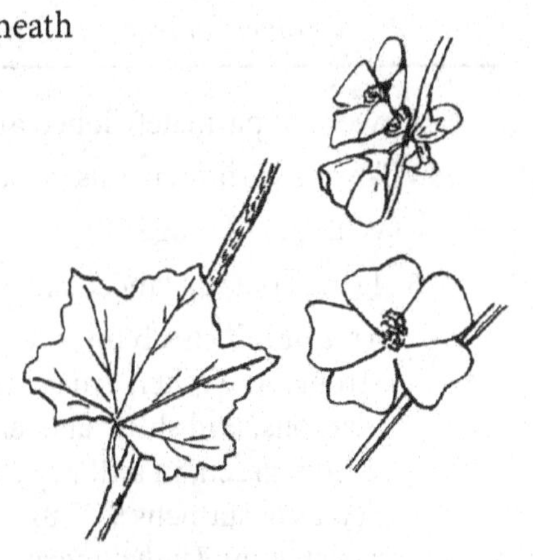

Malvaceae
Malacothamnus fasciculatus (Torr. & A. Gray) Greene

Chaparral Mallow

---

9. Branches pubescent with yellowish hairs. Leaves sparsely pubescent beneath

Erect shrub, 3 to 6 feet high, pubescent with yellowish hairs. Leaves ovate, palmately lobed; ¾ to 2 inches across. Flowers in dense sessile clusters, forming an interrupted spike. Petals 5, pink-rose, ⅜ to ⅝ inch long. Calyx 5-lobed, almost as long as petals. Stamens many, united into a tube around the style. Filiform bractlets, ¼ to ⅝ inch long subtend the flowers. Found on dry slopes, 1,000 to 4,000 feet elevation, from the Santa Ana Mountains to the Cuyamaca Mountains of San Diego.

Malvaceae
Malacothamnus densiflorus (S. Watson) Greene

Yellowstem Bush-Mallow

---

8. Petals fused below. Stamens 5; separate

Shrub 3 to 6 feet high. Young twigs glandular pubescent. Leaves 3 to 5 lobed, lighter green and pubescent beneath, mostly ½ to 1¼ inches long. Flowers rose, in terminal racemes that are 10 to 25 flowered. Corolla salverform, with a tube ¼ inch long. Fruit a dark purple to black berry, ¼ inch in diameter, more or less glandular hairy. Very common shrub. Blooms in early spring, January to March. Berries are edible and make good jelly or pie.

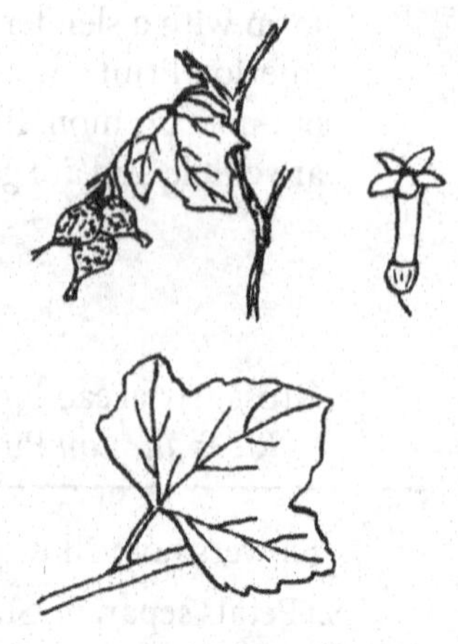

Grossulariaceae
Ribes malvaceum Sm.

Chaparral Currant

---

4. Leaves not palmately lobed and veined
   5. Flowers without petals. Sepals sometimes colored and clustered in a calyx-like involucre (With petals, page 30)
      6. Leaves oblong lanceolate. Herbage strong scented. Flowers small, yellow-green

Tree or tall shrub up to 100 feet tall. Herbage with a strong odor when crushed. Leaves entire, lanceolate, glabrous, and shiny above. Flowers without petals, yellow-green, in axillary clusters on short peduncles. Sepals 6; stamens 9. Fruit a purple drupe, ovoid; about 1 inch long. *Umbellularia* are primitive angiosperms and are grouped within the Magnoliids. Common tree in canyons in the chaparral. More abundant northward. Blooms in early spring, from December to May. Leaves of the California bay were used by the Native Americans to make a tea for headache and stomach pains. The leaves were also used in the houses as a flea repellent.

Lauraceae
Umbellularia californica (Hook. & Arn.) Nutt.

California Bay

---

6. Leaves not lanceolate with strong odor of bay

7. Sepals colored and petal-like; in a cup-shaped involucre
(Sepals green, page 28)

8. Sepals yellow

Low perennial with white woolly leaves. Lower stem woody and covered with old leaves. Leaves ovate, ½ to 1½ inches long. Flowers in a rather dense terminal cluster. Calyx bright yellow. Rare plant, found in crevices on rocky slopes in southern Ventura County on the Conejo Grade. A nice colony of Conejo buckwheat may be observed on the cliffs above the water falls in Wildwood Canyon. Flowers from April to July. **RARE.**

Polygonaceae
Eriogonum crocatum Davidson

Buckwheat Family

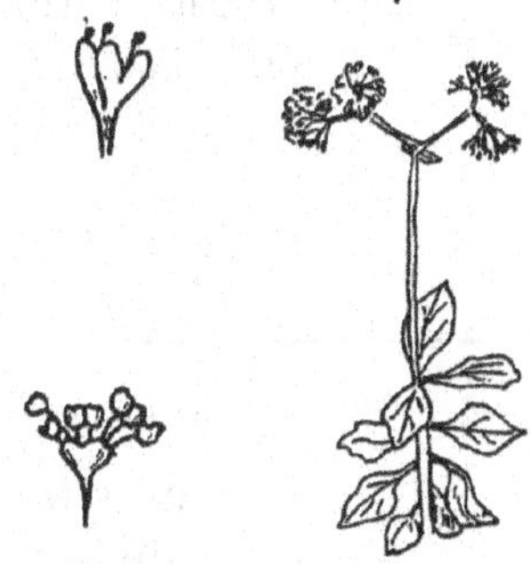

Conejo Buckwheat

8. Sepals pink or white

9. Involucres axillary on an elongated stem

Perennial with erect, few branched stems. Leaves on lower portion of stem; upper stem appearing largely naked. Leaves ovate, undulate, densely white, tomentose beneath; green above. Inflorescence leafless, 2 feet or more long, with axillary involucres. Calyx pinkish-white; 6-lobed. Petals none. Common in dry rocky areas. Blooms in late summer, August to November.

Polygonaceae
Eriogonum elongatum Benth.

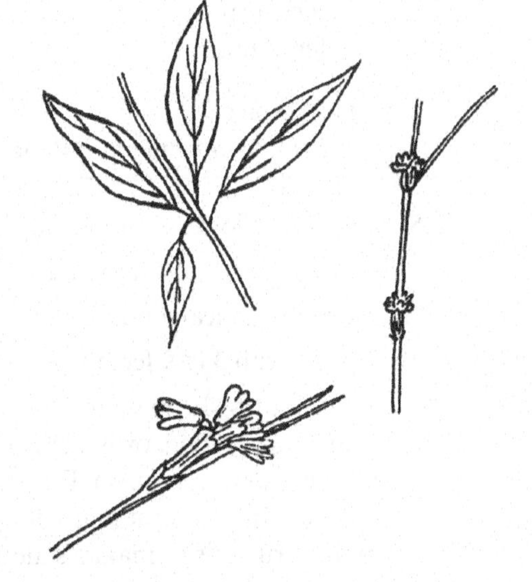

Long-Stem Wild Buckwheat

9. Involucres terminal, forming a compact, ball-like cluster

10. Leaves linear to oblanceolate. Very common

Low shrub with many spreading leafy branches. Leaves linear to oblanceolate, densely fascicled; green above, gray woolly beneath, revolute; ¼ to ⅝ inch long. Flowers pinkish-white in dense terminal heads. Peduncles leafless, 1 to 4 inches long. Very abundant on dry slopes of the chaparral. Blooms from May to October.

Polygonaceae
Eriogonum fasciculatum Benth.

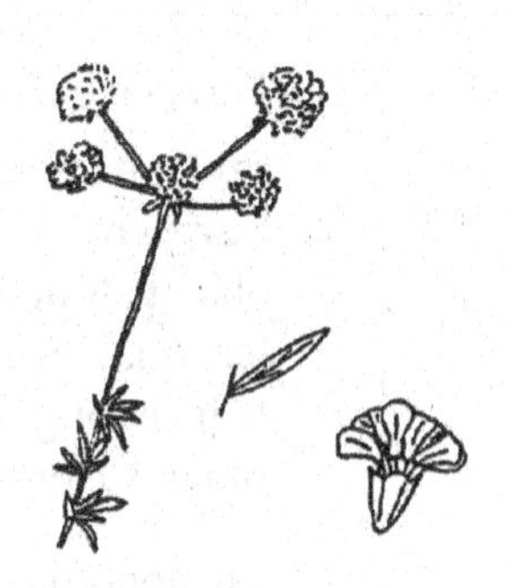

California Buckwheat

10. Leaves ovate

11. Leaves ¼ to ½ inch long. Calyx glabrous

Shrub similar in appearance to the above but with ovate leaves ¼ to ½ inch long. Leaves densely fascicled, green above and densely white tomentose beneath. Involucres and calyx glabrous, white to pink. Plant found only on sand dunes along the beach. Blooms in summer.

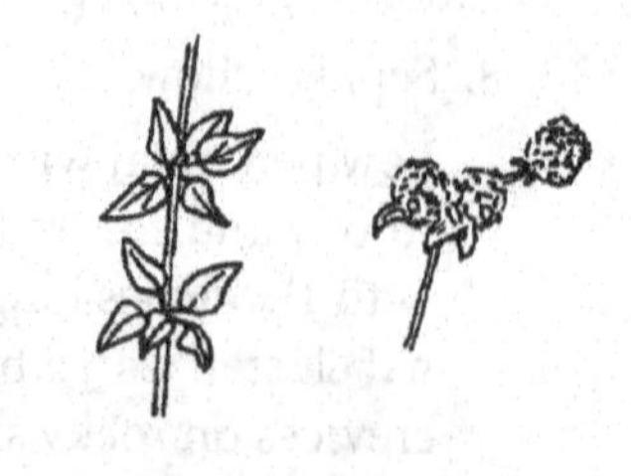

Polygonaceae
Eriogonum parvifolium Sm.

Seacliff Wild Buckwheat

11. Leaves ½ to 1¼ inch long. Involucres and calyx white villous

Low shrub with whitish pubescent stems. Leaves ovate, undulate, green above and white tomentose beneath; ½ to 1¼ inches long. Calyx white to pink. Involucres and calyx white villous. Found on ranges near or facing the coast. Blooms from June to December.

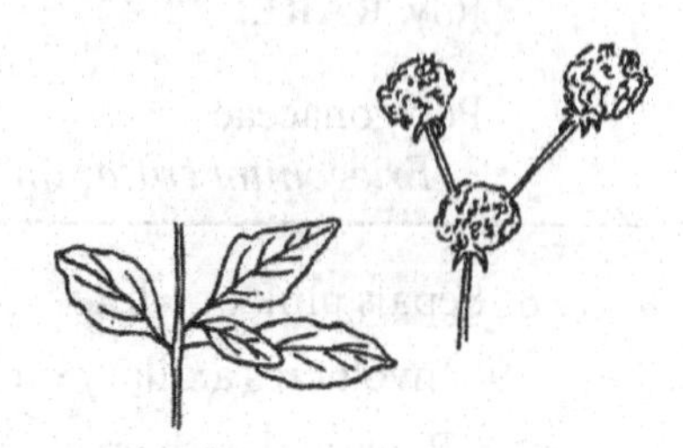

Polygonaceae
Eriogonum cinereum Benth.

Coastal Wild Buckwheat

7. Sepals green. Flowers not in a cup-shaped involucre

8. Plants of beach and salt marsh

9. Leaves gray, mealy; not linear or fleshy

10. Erect shrub 3 to 9 feet tall

Gray, scurfy shrub up to 9 feet tall, dioecious. Leaves 1 to 2 inches long, ovate, entire; sometimes deltoid. Flowers small, greenish. Fruits forming flattened disks, ¼ inch long at maturity; round-ovate. Common shrub of salt marshes along the coast. Blooms in late summer, July to October.

Chenopodiaceae
Atriplex lentiformis (Torr.) S. Watson

Big Saltbush

10. Low spreading shrub

Prostrate perennial, scurfy or mealy when young. Leaves numerous, mostly less than 1 inch long, oblong, toothed, or entire. Flowers in axils and terminal. Fruits slightly fleshy, often red at maturity; ¼ inch long or less; rhombic. Weed of alkaline soils. Blooms April to December.

Chenopodiaceae
Atriplex semibaccata R. Br.

Australian Saltbush

9. Leaves pubescent; linear and fleshy

Perennial plant 1 to 3 feet high. Leaves are linear, pubescent, somewhat fleshy, ½ to 1½ inches long and crowded on the stem. Flowers are small and greenish. Occurs on coastal salt marshes from Santa Barbara to San Diego and blooms from July to October.

Chenopodiaceae
Suaeda taxifolia (Standl.) Standl.

Woolly Seablite

---

8. Plants not of beach and salt marsh

9. Herbage gray, mealy. Low bush in alkaline habitat

Prostrate perennial, scurfy or mealy when young. Leaves numerous, mostly less than 1 inch long, oblong, toothed or entire. Flowers in the axils and terminal. Fruits slightly fleshy, often red at maturity; ¼ inch long or less; rhombic. Weed of alkaline soils. Blooms from April to December.

Chenopodiaceae
Atriplex semibaccata R. Br.

Australian Saltbush

---

9. Herbage not gray, mealy

10. Leaves serrate above the middle; whitish pubescent on under surface. Fruit an achene with a villous tail

Shrub 6 to 20 feet tall. Leaves oval, ½ to 1 inch long, entire below the middle, serrate above; whitish pubescent beneath. Flowers in clusters of 2 or 3. Pistil 1. Fruit an achene with a prominent elongate villous tail. One of the dominant shrubs of the chaparral. Blooms between March and May. Fruits visible during the late summer and fall.

Rosaceae
Cercocarpus betuloides Nutt.

Birch-Leaf Mountain-Mahogany

---

10. Leaves serrate to base; green and mostly glabrous on undersurface. Fruit a red berry

Shrub 3 to 6 feet tall. Leaves glabrous and shining, ovate-elliptic, ½ to 1 inch long; serrate. Flowers greenish inconspicuous. Fruit a red berry. Common dominant shrub of the chaparral. Blooms from March to April.

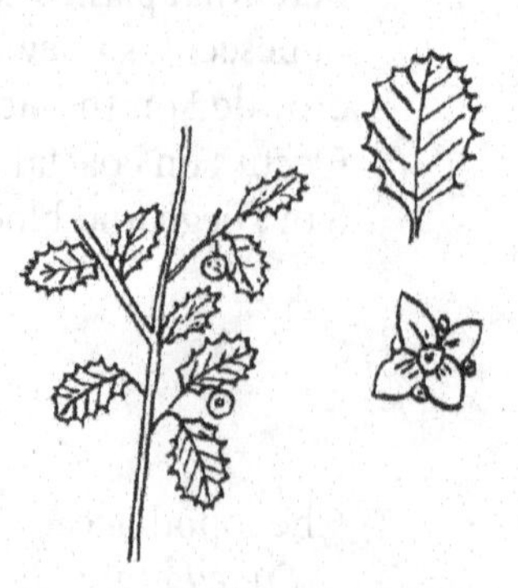

Rhamnaceae
Rhamnus crocea Nutt.

Spiny Redberry

Also found in the area is *Rhamnus ilicifolia*. It is similar to the above species except for the leaves which are spinosely serrate, and ¾ to 1½ inches long. Flowers from March to June.

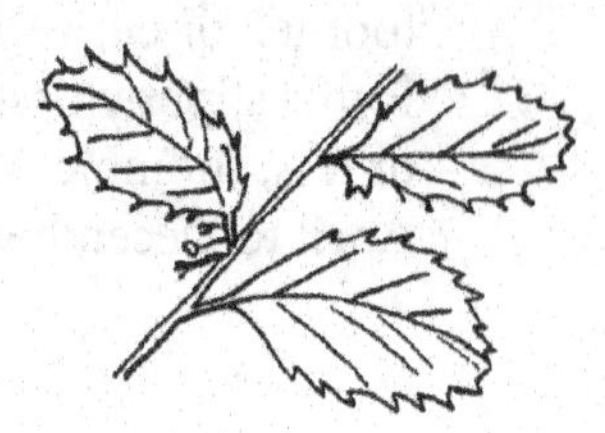

Rhamnaceae
Rhamnus ilicifolia Kellogg

Hollyleaf Redberry

---

5. Flowers with petals
 6. Flowers regular (Flowers irregular, page 39)
  7. Petals large, 1½ to 3 inches long

Glabrous perennial 3 to 8 feet high with a woody base and branched stems. Leaves alternate, pinnatifid, grayish, 2 to 8 inches long. Flowers showy, white. Petals 6, 1½ to 3 inches long. Sepals 3, bristly, hairy. Stamens numerous. Chaparral plant often cultivated for its showy flowers. Blooms from May to July. **Coulter's Matilija Poppy**, *Romneya coulteri* Harv., also occurs in the chaparral from the Santa Ana Mountains to San Diego. It differs in that the sepals are glabrous and the petals are larger, 2 to 4 inches long.

Papaveraceae
Romneya trichocalyx Eastw.

Hairy Matilija Poppy

---

7. Petals smaller, less than 1½ inches long

8. Petals distinct; not fused at the base (Petals fused, page 36)

9. Petals 5 (Petals 4, page 35)

10. Leaves linear

11. Leaves fascicled

Shrub 2 to 8 feet tall. Leaves linear, fascicled, often resinous; ¼ to ½ inch long. Flowers small white, in dense terminal panicles. Petals 5; 1/16 inch long. Sepals 5. Stamens 10 to 15. Dominant shrub of chaparral. Blooms May to June.

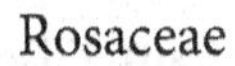

Rosaceae
Adenostoma fasciculatum Hook. & Arn.

Chamise

---

11. Leaves not fascicled

Shrub or small tree 6 to 18 feet tall. Bark reddish brown, commonly peeling in ribbon-like strips. Leaves linear alternate, resinous; ¼ to ½ inch long. Flowers like chamise (above). Chaparral shrub that is generally less abundant than *Adenostoma fasciculatum*. It is locally abundant in the Santa Monica Mountains and on Mt. San Jacinto. Blooms from July to August.

Rosaceae
Adenostoma sparsifolium Torr.

Red Shank

---

10. Leaves not linear

11. Leaves toothed (Leaves entire, page 34)

12. Flowers blue. Fruit a 3-lobed capsule

Shrub 3 to 9 feet tall with densely villous young branches and petioles. Leaves ovate to elliptical, ½ to 1½ inches long with 3 main veins from the base. Flowers in terminal clusters ½ to 2 inches long; deep blue. Petals and sepals 5; stamens 5. Fruit a capsule without horns. Common chaparral shrub, blooming in the early spring, February to April.

Rhamnaceae
Ceanothus oliganthus Nutt.

Hairy Ceanothus

---

12. Flowers not blue

13. Flowers green

Shrub, 3 to 12 feet tall. Leaves oblong to elliptic, revolute, shining above, paler or even silvery pubescent beneath; 1 to 3 inches long, toothed to entire. Flowers in many-flowered umbels, greenish, with 5 petals and sepals; ⅛ inch across. Fruit a berry; green to black. A common chaparral shrub, blooming from May to July.

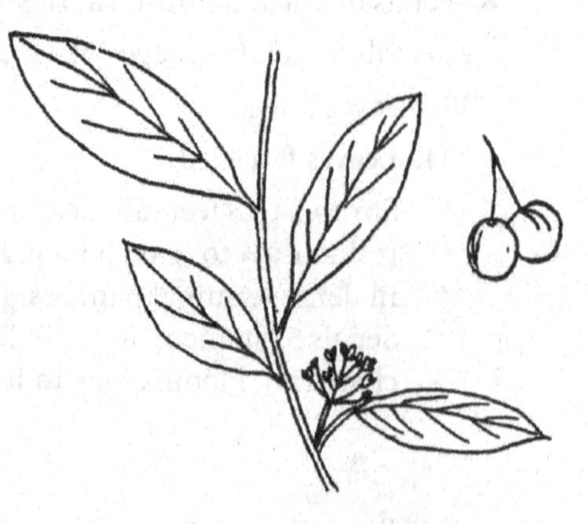

Rhamnaceae
Frangula californica (Eschsch.) A. Gray

California Coffee Berry

---

13. Flowers white or pink

14. Leaves soft, hairy beneath

Shrub 3 to 18 feet tall. Leaves ovate, 1 to 2½ inches long, deeply toothed. Flowers white to pinkish. Petals 5; ¹⁄₁₆ inch long. Stamens 20. Pistils 5. Flowers in a dense terminal inflorescence. Chaparral shrub of the coast ranges, but not too common in Ventura County. Blooms from May to July.

Rosaceae
Holodiscus discolor (Pursh) Maxim.

Oceanspray

---

14. Leaves glabrous beneath

15. Leaves elliptical to orbicular. Apex obtuse or rounded

Shrub 3 to 9 feet tall. Herbage aromatic when crushed. Leaves leathery, oblong-ovate, 1 to 2 inches long, rounded to obtuse at both ends; few toothed or entire. Flowers in short, compact panicles with stout branchlets. Petals 5; white to rose, ⅛ inch long. Sepals 5. Stamens 5. Fruit an acid, sticky pubescent drupe about ⅜ inch in diameter. Blooms in early spring, February to May. The sticky berries may be dropped in water for a lemonade-like beverage. Add sugar or honey to sweeten to taste.

Anacardiaceae
Rhus integrifolia (Nutt.) Rothr.

Lemonade Berry

---

15. Leaves ovate to oblong-lanceolate. Apex pointed

16. Leaves ovate

17. Leaves ¾ to 2 inches long; spinose toothed

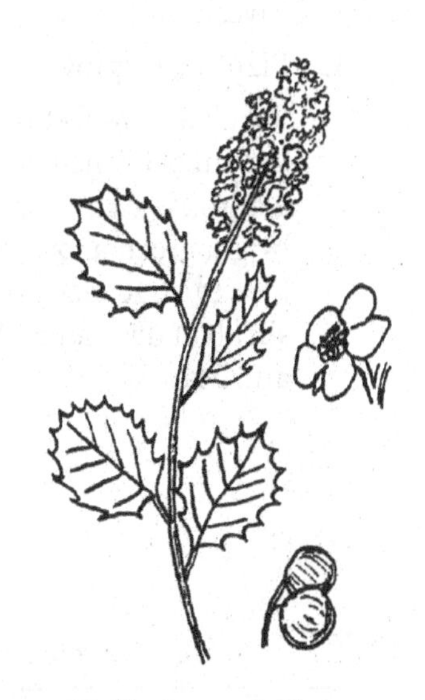

Tree or shrub 3 to 25 feet tall. Leaves leathery, ¾ to 2 inches long, spinose toothed, glabrous and shiny. Flowers small, white, 1 to 2½ inches long in dense terminal racemes. Petals 5; ⅛ inch long. Sepals 5; stamens many; pistil 1. Fruit a red, ovoid drupe, about ½ inch long. Common shrub on dry slopes. Blooms from April to May. Fruits of this species are edible but quite sour. They are best if made into a jelly or jam. **Pits contain cyanide and should not be swallowed**. Cooking the fruits will remove the cyanide as it is volatile.

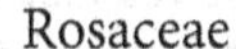

Rosaceae
Prunus ilicifolia (Hook. & Arn.) D. Dietr.

Holly-Leaved Cherry

---

17. Leaves 1½ to 3 inches long; few-toothed to entire

Shrub 5 to 15 feet tall. Young stems and petioles a dark red. Leaves thick and leathery, ovate; entire or few-toothed, sharply acute; 1½ to 3 inches long, more or less folded along the midrib and appearing trough-like. Flowers pinkish-white in dense panicles. Panicle branchlets stout. Petals, sepals, and stamens 5. Fruit reddish, pubescent; sugar-coated; ¼ inch in diameter. Common chaparral shrub, blooming in the spring from March to May. The acid fruits may be dropped in water for a lemonade-like beverage.

Anacardiaceae
Rhus ovata S. Watson

Sugar Bush

---

16. Leaves oblong-lanceolate

Evergreen shrub 6 to 30 feet tall. Leaves oblong, 2 to 4 inches long, sharply toothed, dark green above, lighter beneath, flowers small, white, in terminal clusters. Petals 5; 3/16 inch long. Sepals 5. Stamens 10. Fruit a berry about ¼ inch across. Common chaparral shrub. Blooms in June and July. Berries are red in the late fall. The berries are edible, if cooked or boiled.

Rosaceae
Heteromeles arbutifolia (Lindl.) M. Roem.

Toyon

---

11. Leaves entire

12. Plant with spiny branchlets

Shrub 6 to 18 feet tall with smooth, olive green bark. Terminal branchlets rigid and spinescent. Leaves entire, glabrous, with 1 main vein from the base, elliptic, bright green above; ½ to 1¼ inches long. Flowers pale blue in large clusters, 1½ to 6 inches long. Common shrub of dry slopes, blooming between February and May.

Rhamnaceae
Ceanothus spinosus Nutt.

Greenbark Ceanothus

---

12. Plant without spiny branchlets

13. Leaves with 3 main veins from base. Flowers blue

Shrub 3 to 12 feet tall with greenish branches. Leaves 3-veined from the base, deciduous, ovate, 1 to 3 inches long. Flowers small, white to pale blue, in showy panicles. Petals 5; stamens 5. Common on dry slopes. Blooms from May to June.

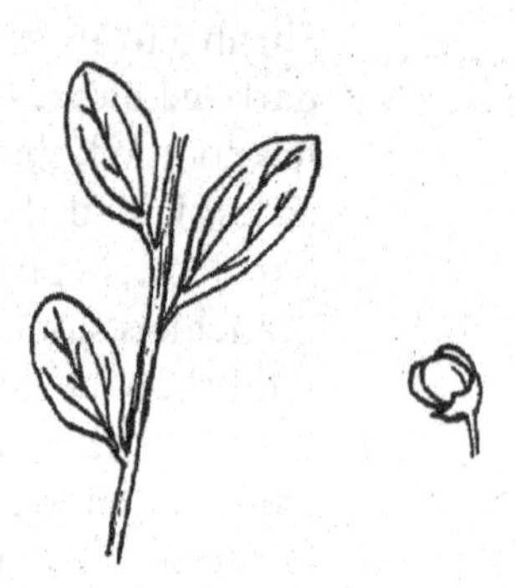

Rhamnaceae
Ceanothus integerrimus Hook. & Arn.

Deer Brush

---

13. Leaves with 1 main vein from base

14. Leaves small, less than 1 inch long. Fruit a 3-lobed capsule

Shrub 3 to 12 feet tall. Leaves leathery, spathulate, truncate or notched at the apex, entire, glabrous above, evergreen, slightly revolute; ¼ to ¾ inch long. Flowers white in umbels on lateral branchlets. Common chaparral shrub, particularly near the coast. Blooms from January to April.

Rhamnaceae
Ceanothus megacarpus Nutt.

Bigpod Ceanothus

---

14. Leaves 1 to 4 inches long. Fruit not a 3-lobed capsule

15. Leaves somewhat folded along midrib

16. Leaves ovate

See *Rhus ovata*, page 33.

16. Leaves oblong to lanceolate

Evergreen shrub 6 to 15 feet tall; glabrous and aromatic. Young stems and petioles dark red. Leaves lanceolate to oblong, entire, 1¼ to 4 inches long, slightly folded along the midrib and with a mucronate tip. Flowers tiny, white, in dense terminal inflorescences. Fruit white, glabrous. A common shrub more commonly found in the wooded valley flats. Blooms during the summer months, June and July.

Anacardiaceae
Malosma laurina (Nutt.) Abrams

Laurel Sumac

---

15. Leaves flat, not folded along midrib

16. Leaves elliptical to orbicular. Fruit a sticky acid drupe

See *Rhus integrifolia*, page 32.

16. Leaves oblong-lanceolate. Fruit a berry

See *Frangula californica*, page 32.

9. Petals 4

10. Flowers yellow

11. Shrub. Leaves lanceolate, entire

Shrub 3 to 10 feet tall. Leaves entire, lanceolate, 1 to 4 inches long. Flowers showy with 4 petals averaging 1 inch long; 2 sepals; numerous stamens. Common chaparral shrub, especially after fires. Blooms from April to June.

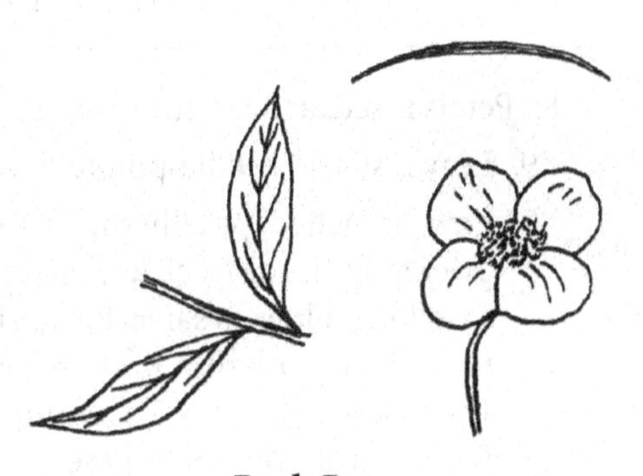

Papaveraceae
Dendromecon rigida Benth.

Bush Poppy

---

11. Perennial. Lower leaves pinnatifid

Perennial, 1¼ to 5 feet tall. Lower leaves pinnatifid, upper leaves divided or entire. Flowers yellow with 4 petals ½ to ¾ inch long. Fruit a long linear pod 1¼ to 3 inches long, spreading out from the main stem. Desert perennial often found in sandy soils. This plant tolerates selenium in the soil and has been used as a selenium indicator. Blooms between April and September. It is not common in our area, although it can be found on the coast ranges of the Santa Monica Mountains about 100 feet up from the coast highway near Malibu pier.

Brassicaceae
Stanleya pinnata (Pursh) Britton

Prince's Plume

---

10. Flowers orange to red

Perennial with stems 1 to 3 feet long. Slightly shrubby at the base. Leaves linear to ovate lanceolate; toothed or entire. Flowers red; petals and sepals 4; both red. Stamens 8; 4 long and 4 short. Corolla base tubular. Ovary inferior. Common on dry stony spots. There are two subspecies commonly found. **Hoary Fuchsia**, *Epilobium canum* (Greene) P.H. Raven subsp. *canum*, has narrower fascicled leaves that are usually grayer. Both species bloom in late summer from August to October.

Onagraceae
Epilobium canum (Greene) P.H. Raven
subsp. *latifolium* (Hook.) P.H. Raven

California Fuchsia

---

8. Petals fused, at least at the base

9. Leaves sharp, needle-pointed; fascicled. Flowers showy, pink or blue

Low, branched, prickly shrub 1 to 3½ feet high. Leaves prickly, in dense fascicles, linear and needlelike; ⅛ to ½ inch long. Flowers salverform, showy, in terminal clusters. Corolla 5-lobed, pink to lavender to white; up to 1 inch long. Stamens 5. Common shrub on dry slopes. Blooms from March to June.

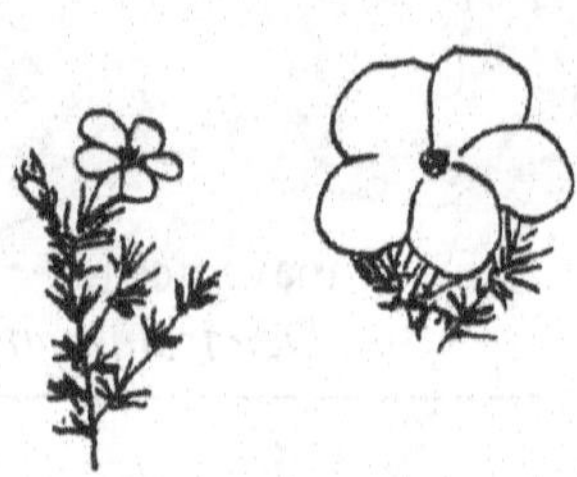

Polemoniaceae
Linanthus californicus (Hook. & Arn.)
J.M. Porter & L.A. Johnson

Prickly Phlox

---

9. Leaves not sharp, needlelike

10. Flowers yellow; long tubular

Shrub or small tree, 6 to 25 feet tall with drooping branches. Leaves glabrous, entire, ovate or ovate-lanceolate; 1 to 3 inches long; long petioled. Flowers pale yellow, long tubular, contracted at the throat; about 1½ inches long. Common shrub in waste places. Blooms during spring and summer.

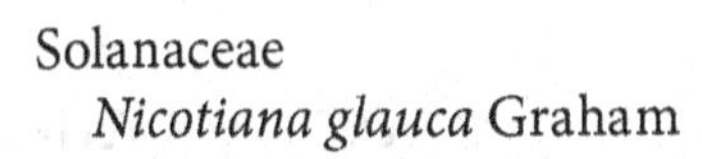

Solanaceae
Nicotiana glauca Graham

Tree Tobacco

---

10. Flowers not yellow

11. Corolla rotate. Stamens exserted

12. Flowers white. Corolla deeply 5 cleft

Perennial 3 to 6 feet tall. Stem angled and slightly pubescent. Leaves ovate, sinuate dentate; ¾ to 4 inches long. Flowers in clusters on peduncles up to 1 inch or more long. Corolla white, rotate, deeply 5 cleft, up to ½ inch long (if corolla clefts less than ⅛ inch, see **Small-flowered Nightshade**, *Solanum americanum* Mill.). Stamens 5; anthers yellow, in a tube around the pistil. Fruit a black berry. **POISONOUS** to humans. Common plant of the chaparral or wooded areas. Blooms most of the year.

Solanaceae
Solanum douglasii Dunal

White Nightshade

---

12. Flowers purple to blue. Corolla shallow lobed

Glandular pubescent perennial with a woody base. Leaves ovate, ¾ to 1½ inches long, mostly entire or slightly sinuate dentate. Flowers in lateral umbels, deep blue to violet. Corolla up to 1 inch across, shallow lobed and slightly angled. Anthers yellow. Berry greenish, ¼ inch in diameter. Common chaparral perennial. Blooms from May to July. The berries are **POISONOUS** and should not be eaten.

Solanaceae
Solanum xanti A. Gray

Purple Nightshade

---

11. Corolla not rotate. Stamens not exerted

12. Leaves toothed. Flowers lavender

Shrub 3 to 9 feet tall. Leaves, twigs, and calyx gray or brownish tomentose. Leaves lanceolate-ovate, 2 to 6 inches long, crenate or coarsely dentate. Flowers in terminal coiled inflorescences. Corolla funnelform, about ½ inch long, pale lavender. Common chaparral shrub, blooming from April to June.

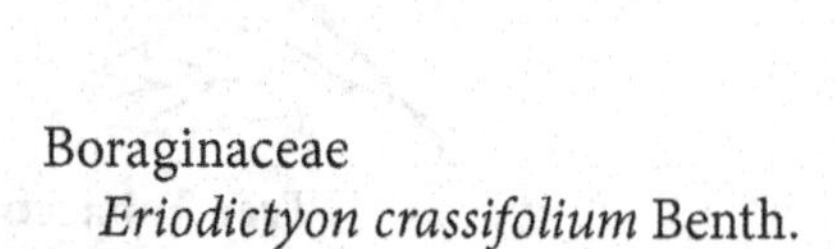

Boraginaceae
Eriodictyon crassifolium Benth.

Thick-Leaved Yerba Santa

---

12. Leaves entire. Flowers pinkish-white; urn-shaped

13. Branchlets glabrous

Woody evergreen shrub or small tree with crooked branches and smooth red, peeling bark. Leaves simple, glabrous, elliptic, ¾ to 1¾ inches long. Flowers in dense panicles, nodding. Corolla white or pinkish, urn-shaped, ⅛ to ⅜ inch long; 5-lobed with the lobes recurved. Fruit a globose sticky berry, about ½ inch across. There are many species of manzanita. This is the most common species in the Santa Monica Mountains. Manzanita usually occurs in moister areas on the coast side of mountains at high elevations. Blooms between December and March. The berries are quite acid but make excellent jellies and pie.

Ericaceae
Arctostaphylos glauca Lindl.

Bigberry Manzanita

---

13. Branchlets pubescent or covered with woolly hairs

Erect shrub 4 to 8 feet high with smooth, reddish stems. Young branchlets are glandular hairy. Leaves dull green, ovate to lanceolate-ovate, 1 to 1¾ inches long and ½ to 1 inch wide. Petioles ¼ inch long, glandular hairy. Corolla white, urn-shaped or bell-shaped, 3/16 to ⅝ inch long. Fruit a globose, sticky berry, 5/16 inch across. Crown manzanita grows in gravelly places above 1,000 feet, from San Diego County north. **Del Mar Manzanita,** *A. glandulosa* Eastw. subsp. *crassifolia* (Jeps.) P.V. Wells, occurs along the coast in San Diego County. It differs in being only 2 to 4 feet high. Leaves are dark green above and woolly beneath. Branchlets are woolly, but not glandular.

Ericaceae
Arctostaphylos glandulosa Eastw.

Eastwood Manzanita

6. Flowers irregular
   7. Flowers yellow, papilionaceous. Leaves few to none

      Tall mostly leafless shrub up to 9 feet tall. Leaves, when present, small, entire, oblanceolate-oblong; ½ to 1 inch long. Flowers large, fragrant, yellow; in terminal racemes. Petals ¾ to 1 inch long. Fruit a linear pod 2 to 4 inches long. Blooms from April to June. This plant is often cultivated and frequently occurs as an escape. It is rapidly becoming naturalized, particularly along the coast and at higher elevations in the mountains.

      Fabaceae
      Spartium junceum L.

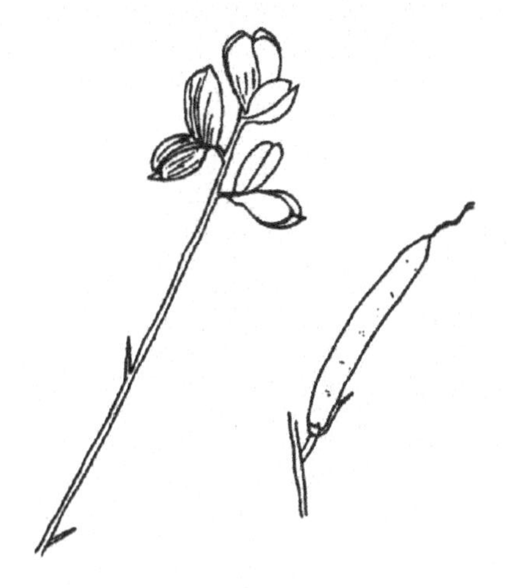

Spanish Broom

---

   7. Flowers green, but appearing red because of red-tipped calyx and bracts. Leaves linear, fascicled

      Perennial with grayish herbage, forming a bushy growth 1 to 2 feet high. Leaves are white-woolly, oblong to linear, and entire or 3-lobed above. Flowers are greenish but appear red because of scarlet-tipped bracts and calyxes. Occurs on dry open places and blooms from March to June.

      Orobanchaceae
      Castilleja foliolosa Hook. & Arn.

Woolly Paintbrush

SECTION II
HERBACEOUS DICOTS

1. Plants without petals, or petals extremely minute; less than ⅛ inch long. Sepals may or may not be present. Flowers mostly greenish, but sepals or subtending bracts sometimes colored and corolla-like.

(Plants with petals, page 63)

2. Plants found only on dunes and in salt marshes along the coast

(Plants not on dunes and in salt marshes along the coast, page 47)

3. Leaves absent. Plants jointed and succulent

Succulent, leafless plant with jointed stems. Branches opposite. Flowers minute, sunken in the upper joints. Common plant of salt marshes. Blooms in late summer from August to November.

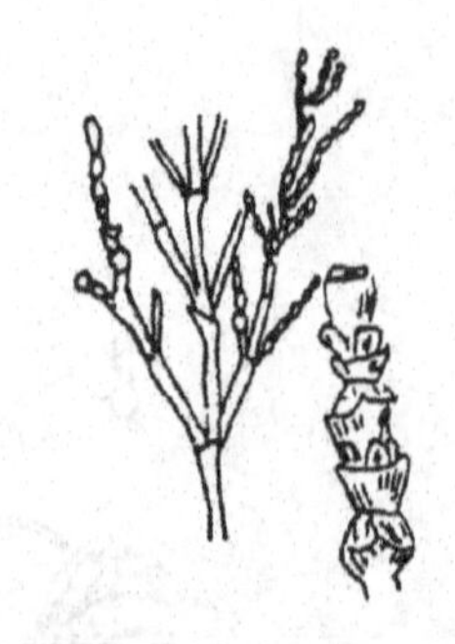

Chenopodiaceae
Salicornia pacifica Standl.

Pacific Pickleweed

---

3. Leaves present. Stem not jointed

4. Leaves pinnatifid dissected. Fruit a sharp, prickly bur

Low spreading plant, forming mats 3 to 9 feet across and 6 to 12 inches high. Leaves silvery, pinnatifid dissected. Flowers unisexual in composite heads. Heads in terminal spikes. Male heads bowl-shaped; female heads with sharp spines, becoming a bur in fruit. Common plant on sand dunes along the coast. Blooms from March to September.

Asteraceae
Ambrosia chamissonis (Less.) Greene

Beach Bur-Sage

---

4. Leaves not pinnatifid dissected

5. Leaves sheathing with scarious stipules

Glabrous perennial with unbranched stems. Leaves lanceolate, 2 to 4 inches long, with scarious sheathing stipules. Flowers in a terminal spike. Calyx rose, ⅜ inch long. Common plant of marshy areas.

Polygonaceae
Persicaria amphibia (L.) Gray

Water Smartweed

---

5. Leaves not sheathing with scarious stipules
  6. Leaves mostly basal
    7. Flowers in a dense, terminal spike; subtended by petal-like bracts

Flowering stem with one clasping leaf in middle of stem and 1 to 3 petioled leaves in the axil. All other leaves basal, long petioled (often 4 to 6 inches long) and elliptic or oblong. Leaf blade large, 4 to 5 inches long. Stem and petioles covered with fine curled hairs. Flowers small, in dense terminal spike, subtended by 5 to 8 whitish or rose streaked bracts. Floral spikes usually ½ to 1½ inches long. *Anemopsis* are primitive angiosperms and are grouped within the Magnoliids. Common plant of wet alkaline places. Blooms from March to September.

Saururaceae
Anemopsis californica (Nutt.) Hook. & Arn.

Yerba Mansa

---

7. Flowers in a small globose head; subtended by short woolly bracts

Prostrate annual with reddish, glabrous stems 4 to 12 inches long. Leaves are mostly basal, occurring in a tuft. Leaves are white woolly; spathulate; ¾ to 2 inches long. Leaves on main stem are much reduced. Flowers are small, yellowish or pinkish and occur in a head. Found on beaches and dunes along the coast from Los Angeles County south to San Diego. Blooms from April to September.

Polygonaceae
Nemacaulis denudata Nutt.

Coast Woolly-Heads

---

6. Leaves not mostly basal
  7. Calyx colored and corolla-like; clustered in an involucre
    8. Calyx salverform; showy; ½ inch long; pink to red-violet
      9. Flowers red-violet. Leaves oval; thick and leathery

Prostrate perennial with glandular pubescent herbage. Stems thick and distinctly woody near base. Leaves oval, entire, with rounded apex; thick and leathery. Flowers salverform, 5-lobed, with each lobe shallowly 2-lobed. Flowers in a head-like cluster subtended by an involucre of 5 to 8 leaflike bracts. Calyx rose-violet with a delicate fragrance. Common plant of sand dunes. Blooms from February to October.

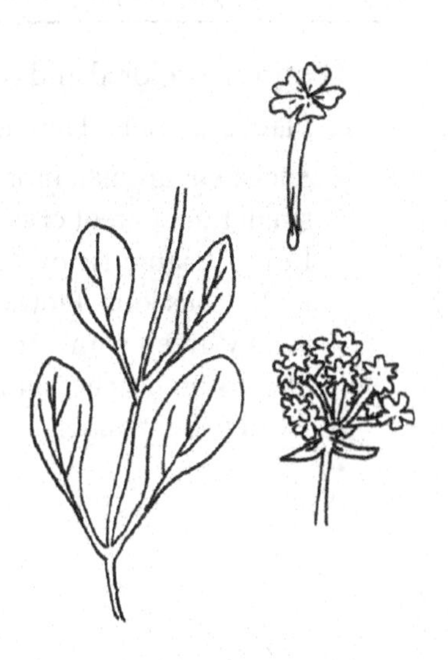

Nyctaginaceae
Abronia maritima S. Watson

Red Sand-Verbena

---

9. Flowers pink. Leaves ovate; not thick and leathery

Prostrate perennial with glandular pubescence and appearance similar to the above. Stem and leaves are not thickened. Leaves ovate, entire or crenulate. Apex acute to obtuse. Flowers similar to above except that they are lighter in color. Also common on sand dunes along the coast. Blooms most of the year.

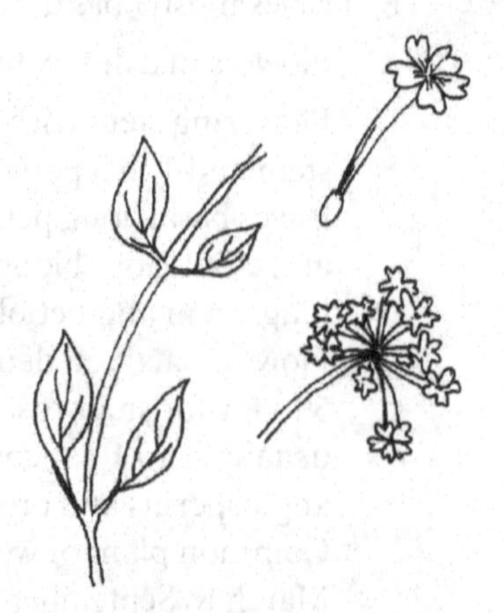

Nyctaginaceae
Abronia umbellata Lam.

Pink Sand-Verbena

---

8. Calyx rotate; ⅛ inch long; pinkish white. Flowers forming ball-like clusters at tips of branches

Low shrub, woody at the base, with small ovate, fascicled leaves. Leaves white tomentose beneath. Flowers small, 6-parted, pinkish white with 9 stamens. Flowers clustered in a small cylindrical involucre and forming a ball-like cluster at the tips of branches. Common shrub of sand dunes. Blooms during summer.

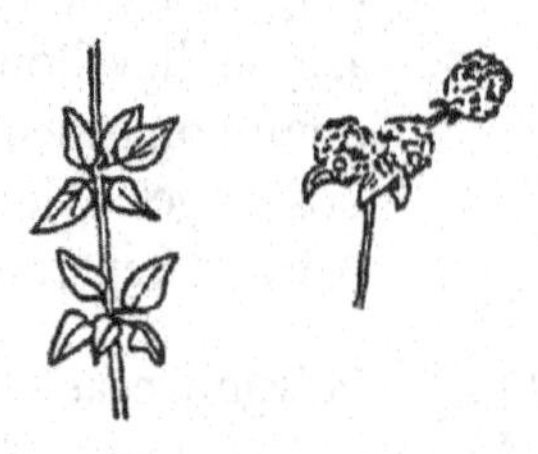

Polygonaceae
Eriogonum parvifolium Sm.

Seacliff Wild Buckwheat

---

7. Calyx not colored and corolla-like. Flowers small, greenish
8. Plant succulent. Herbage with small crystalline vesicles

Succulent annual, more or less prostrate. Herbage with small transparent crystalline vesicles filled with fluid. Leaves triangular ovate, fleshy, entire, undulate; ¾ to 2 inches long. Flowers small, solitary in the axils; yellow-green with 3 to 5 sepals. Fruit a horned capsule. Occurs on beaches and in salt marshes. Blooms from April to September.

Aizoaceae
Tetragonia tetragonioides (Pall.) Kuntze

New Zealand Spinach

---

8. Plant not succulent. Herbage without crystalline vesicles

9. Fruits enclosed in small, disklike bracts. Plants mealy or gray scurfy

10. Fruits slightly swollen; red and fleshy at maturity

Prostrate perennial; woody at the base and scurfy or mealy when young. Leaves small and numerous, averaging ½ to 1 inch long, dentate to entire. Flowers in leaf axils. Fruits red at maturity, rhombic, about ⅛ inch long. Common plant in alkaline soils. Blooms from April to December.

Chenopodiaceae
Atriplex semibaccata R. Br.

Australian Saltbush

---

10. Fruits flattened; gray to green

11. Plant erect. Leaves triangular

Erect, glabrous or slightly mealy plant. Leaves triangular or oval hastate. Flowers unisexual. Fruiting bracts green, rhombic oval. Plant of coastal salt marshes. Not common. Blooms from June to November.

Chenopodiaceae
Atriplex prostrata DC.

Fat-Hen

---

11. Plant prostrate. Leaves round-ovate

Prostrate plant with white scurfy stem and leaves. Woody at base. Leaves numerous, sessile, round-ovate; ½ to 1½ inches long with rounded apex. Fruits ovate, ¼ inch long, somewhat spongy with warty projections. Inhabitant of sandy areas along the coast. Blooms from April to October.

Chenopodiaceae
Atriplex leucophylla (Moq.) D. Dietr.

Beach Saltbush

---

9. Fruits not enclosed in small disk-like bracts

10. Leaves linear. Plant hairy or pubescent

11. Calyx villous to woolly; lobes with a spine on the back. Plant densely hairy

Gray pubescent annual, prostrate or erect, with stems branched from the base. Leaves alternate, entire, linear to linear-lanceolate ¾ to 1½ inches long. Flowers minute in the leaf axils; soft pubescent or woolly. Common weed of alkaline soils; abundant in salt marshes along the coast. Blooms from July to October.

Chenopodiaceae
Bassia hyssopifolia (Pall.) Kuntze

Five-Hook Bassia

---

11. Calyx pubescent; lobes without a spine on the back. Plant pubescent

Perennial 1 to 3 feet high. Leaves are linear, pubescent, somewhat fleshy, ½ to 1½ inches long and crowded on stem. Flowers are small and greenish. Occurs on coastal salt marshes from Santa Barbara to San Diego and blooms from July to October.

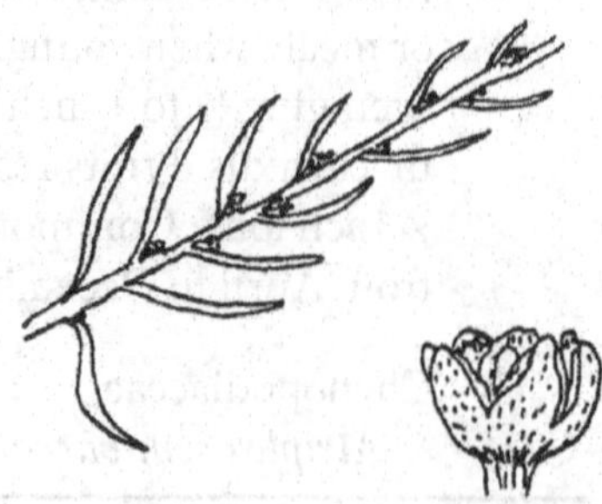

Chenopodiaceae
Suaeda taxifolia (Standl.) Standl. **Woolly Seablite**

---

10. Leaves ovate to oblong. Plant mealy to white hoary; not pubescent

11. Leaves ovate, sinuate dentate. Plant slightly mealy

Stout, erect annual 4 to 20 inches high. Leaves ovate, sinuate dentate; ½ to 2 inches long. Flowers in dense axillary and terminal clusters subtended by leafy bracts. Flowers greenish, fleshy, often becoming reddish at maturity. Inhabitant of moist areas along the coast. Blooms from July to October.

Chenopodiaceae
Chenopodium macrospermum Hook f. **Coast Goosefoot**

---

11. Leaves oblong, entire. Plant white hoary

Erect perennial with alternate, entire leaves. Stem and leaves grayish white scurfy or hoary. Leaves elliptic to oblong, entire, ¾ to 2 inches long. Inhabitant of dry sandy areas, frequently found on sand along the beach. Blooms from March to October.

Euphorbiaceae
Croton californicus Müll. Arg. **California Croton**

---

2. Plants not found only on dunes or in salt marshes along the coast

3. Leaves basal or essentially so. Flowering stem mostly leafless

(Leaves on main flowering stem, page 50)

4. Flowers in a terminal spike or raceme

5. Flowers in a raceme. Fruit a round pod

Pubescent annual 4 to 12 inches high Leaves simple, alternate, mostly all near the base. Leaves ovate to oblong, $^3/_{16}$ to ¾ inch long. Flowers minute in a one-sided raceme. Pedicels slender and curved downward in fruit. Fruits tiny, orbicular with hooked bristles. *Athysanus* is found in grassy or brushy places of the chaparral. Blooms from March to June.

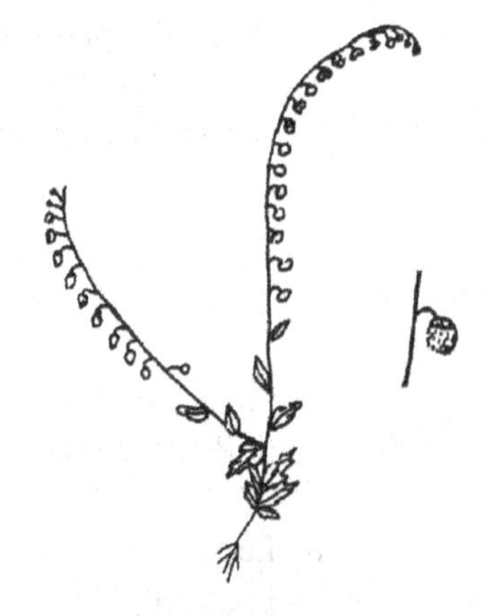

Brassicaceae
Athysanus pusillus (Hook.) Greene

Dwarf Athysanus

---

5. Flowers in a spike

6. Floral spike subtended by petal-like bracts

Flowering stem with one clasping leaf in middle of stem and 1 to 3 petioled leaves in the axil. All other leaves basal, long petioled. Leaf blade elliptic or oblong in shape; 4 to 5 inches long. Flowers small in dense terminal spikes, subtended by 5 to 8 whitish or rose streaked bracts. *Anemopsis* are primitive angiosperms and are grouped within the Magnoliids. Common plant of marshy places. Blooms from March to September.

Saururaceae
Anemopsis californica (Nutt.) Hook. & Arn.

Yerba Mansa

---

6. Floral spike not subtended by petal-like bracts

7. Leaves broadly ovate, 3 to 6 inches long

Perennial plant with basal leaves and acaulescent or leafless stem. Leaves ovate, longitudinally ribbed with conspicuous nerves which converge at base and apex. Flowers in a narrow, dense, terminal spike, 2 to 16 inches long. Flowers 4-merous. Both sepals and petals present, but the petals are quite small and inconspicuous. Corolla lobes $^1/_{25}$ inch; scarious, greenish. Common weed of damp places. Blooms from April to September. Young leaves of common plantain may be used in salads or boiled for a potherb.

Plantaginaceae
Plantago major L.

Common Plantain

---

7. Leaves narrower
   8. Leaves lanceolate

Perennial plant with basal leaves. Leaves lanceolate; blades 2 to 8 inches long. Flowering spike dense, conic to ovoid, ¾ to 3 inches long. Common weed in moist places. Blooms from April to August.

Plantaginaceae
Plantago lanceolata L.

English Plantain

---

8. Leaves linear. Herbage silky pubescent

Low, villous annual with scapes 2 to 10 inches high. Leaves linear to linear lanceolate, 1 to 5 inches long. Flowering spikes short, ¼ to 1 inch long; sepals villous. Corolla lobes ¹⁄₁₆ inch long. Common plant of dry slopes. Blooms from March to May.

Plantaginaceae
Plantago erecta E. Morris

California Plantain

---

4. Flowers not in a terminal spike or raceme
   5. Plant tall (2 to 4 feet), erect, sparingly branched. Flowers, several in an involucre

Perennial with long (2 to 4 feet), mostly leafless stem; whitish tomentose. Leaves mostly basal; stem leafless above the inflorescence. Leaves ovate, undulate, white beneath. Flowers axillary, clustered within an involucre. Calyx pinkish-white; 6-lobed with 9 stamens. Involucre cylindrical, tomentose; ¼ inch long. Common plant of dry rocky areas. Blooms from August to November.

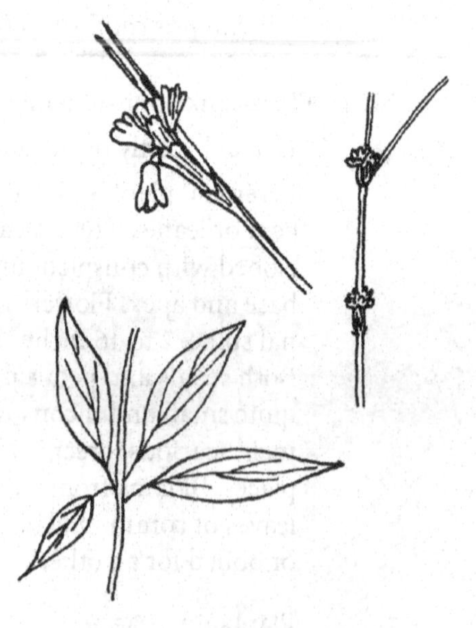

Polygonaceae
Eriogonum elongatum Benth.

Long-Stem Wild Buckwheat

---

5. Plant low (less than 1 foot tall), intricately branched. Involucres 1-flowered

6. Calyx yellow-green

Low annual with trailing, brittle stems 2 to 10 inches long with short hairs. Leaves are mostly basal, oblong to oblanceolate, 1 to 3 inches long. Flowers are yellow-green and occur in axillary and terminal clusters. Found in sandy places from San Fernando Valley south to San Diego. Blooms from April to July.

Polygonaceae
Chorizanthe procumbens Nutt.

Prostrate Spineflower

---

6. Calyx white, pink, or rose

7. Calyx rose colored, without fringed margins

Low, much branched annual (4 to 8 inches high) with reddish more or less pubescent stems. Leaves basal, oblong to obovate; main stem leafless. Each flower in a cylindric, 6-ribbed involucre. Involucres with 6 hooked teeth; ⅛ inch long. Calyx corolla-like; pink to rose; 6-lobed; $^{3}/_{16}$ inch long. Calyx lobes unalike; 3 inner ones smaller. Common plant on dry rocky hills. Blooms from April to June.

Polygonaceae
Chorizanthe staticoides Benth.

Turkish Rugging

---

7. Calyx white-pink with fringed margins

Low, erect annual with reddish pubescent stems 2 to 12 inches high. Leaves are basal; obovate to spathulate; ¾ to 2 inches long. Calyx is white to pink and has fringed margins. Involucral bracts are dark red and have spiny teeth. Flowers occur in a dense, terminal cluster. Found on dry slopes from west Riverside County to west San Diego County. Blooms from April to June.

Polygonaceae
Chorizanthe fimbriata Nutt.

Fringed Spineflower

---

3. Leaves present on main flowering stem; not all basal

4. Leaves compound, pinnately divided or dissected (at least the lower leaves)

5. Flowers in a simple or compound umbel

See **Apiaceae**, page 64.

5. Flowers not in simple or compound umbel

6. Flowers in a raceme. Fruit a linear or rounded pod

See **Brassicaceae**, page 88.

6. Flowers not in a raceme. Leaves 2 to 3 times compound into rounded leaflets

Erect perennial 2 to 6 feet high. Leaves alternate, 3 times compound, glabrous. Flowers are unisexual without petals. Sepals 4, soon withering, petaloid or greenish. Stamens and pistils many, dangling from the sepals. Flowers in a loose panicle. Meadow-rue is fairly common in oak woods or chaparral. Blooms from March to June.

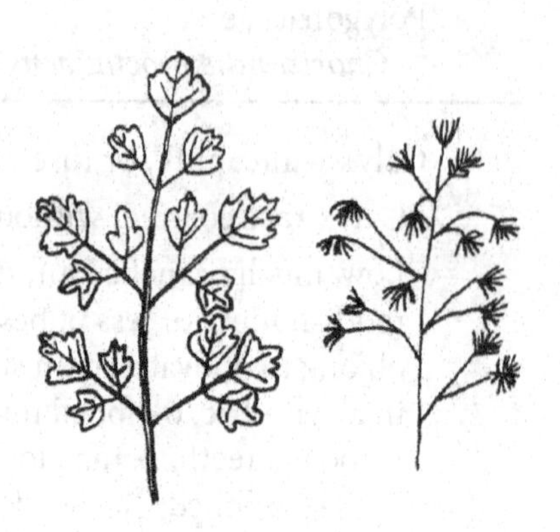

Ranunculaceae
Thalictrum fendleri A. Gray

Meadow-Rue

---

4. Leaves simple

5. Leaves alternate, at least below (Leaves opposite or whorled, page 59)

6. Plants with milky juice

Glabrous, erect annual, 4 to 14 inches high, with milky juice. Stem and leaves alternate, petioled, entire; obovate with rounded apex. Upper leaves opposite. Flowers terminal, greenish. Common garden weed; found in moist places. Blooms from February to August. The milky juice of this species has a **TOXIC** component as it will cause **SEVERE BURNING AND IRRITATION OF THE EYE.** Because petty spurge occurs as a common weed, take care to avoid rubbing the eyes after picking the plant.

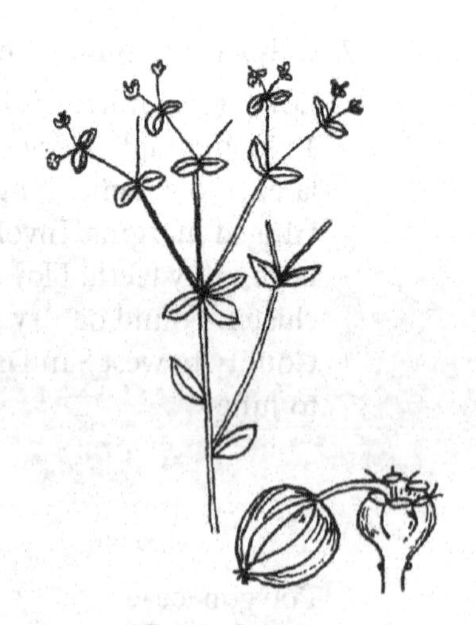

Euphorbiaceae
Euphorbia peplus L.

Petty Spurge

---

6. Plants without milky juice
  7. Leaves with scarious sheaths
    8. Fruit a flattened, papery disk (enlargement of calyx)
      9. Fruiting disks large; ⅜ to ⅝ inch long

Perennial with smooth, reddish stout stems, 2 to 4 feet high. Leaves lanceolateovate; 2½ to 12 inches long; margins crisped. Sheaths pink, scarious ½ to 1¼ inches long. Flowers numerous in a dense, terminal panicle. Calyx of 6 sepals; 3 becoming enlarged in fruit. Fruits averaging ½ inch long; cordateovate. Common plant in dry sandy places. Blooms from January to May.

Polygonaceae
Rumex hymenosepalus Torr. **Wild-Rhubarb**

---

9. Fruiting disks smaller; ¼ inch or less long
10. Flowers in loose, axillary whorls. Leaves undulate, crisped

Perennial with smooth stem, 1½ to 4 feet high. Leaves lanceolate to linear lanceolate; 4 to 12 inches at base; reduced and narrower above. Flowers in a narrow inflorescence of interrupted whorls of flowers. Fruits ¼ inch long or less; round-ovate, with a small wart or callosity on each of the three enlarged calyx valves. Very common weed of waste places. Blooms most of the year.

Polygonaceae
Rumex crispus L. **Curly Dock**

---

10. Flowers in a dense, terminal spikelike panicle. Leaves not undulate, crisped

Perennial with glabrous stems, 1 to 3 feet high. Leaves linear lanceolate, willow-like; 2½ to 5 inches long, flat, entire. Flowers in a dense spike-like panicle. Fruits triangular, about ⅛ inch long; one of the enlarged calyx valves with a large grain or callosity. Inhabitant of moist areas. Less common than the above. Blooms from May to September.

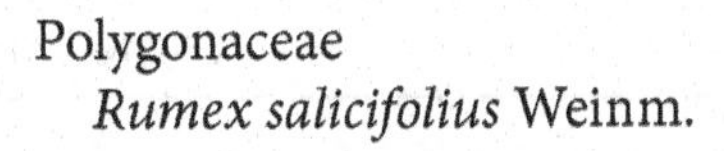

Polygonaceae
Rumex salicifolius Weinm. **Willow Dock**

---

8. Fruits not as above

9. Plant prostrate. Flowers axillary. Common weed of lawn and garden

Prostrate annual with wiry blue-green stems spreading along surface of ground. Leaves oblong to lanceolate, ¼ to ¾ inch long; entire. Stipule sheaths scarious; silvery and torn. Flowers in axils; 1 to 5 flowered. Calyx pinkish, $1/16$ inch long. Very common weed of lawn and waste places. Blooms from May to November.

Polygonaceae
Polygonum aviculare L.

Common Knotweed

---

9. Plant erect

10. Weed of fields

Erect, branched, annual with linear leaves ¼ to 1¼ inches long. Leaf sheaths are silvery and irregularly torn. Calyx is green with rose-colored margins. Flowers occur in small clusters at the nodes. Weed. Blooms from June to October.

Polygonaceae
Polygonum argyrocoleon Kunze

Persian Knotweed

---

10. Plant of water or moist habitat

11. Flowers rose; in a dense-flowered, terminal ovoid spike

Glabrous perennial with unbranched stem. Leaves lanceolate, 2 to 4 inches long, with scarious sheathing stipules. Flowers in an ovoid terminal spike, ½ to 1 inch long. Calyx rose; ⅜ inch long. Common plant of marshy areas.

Polygonaceae
Persicaria amphibia (L.) Gray

Water Smartweed

---

11. Flowers white to greenish; in spike-like racemes

Erect perennial, with stems 1 to 3½ feet high. Leaves are lanceolate, 2 to 4 inches long, glandular punctate. Flowers greenish white in spike-like racemes, forming an elongate panicle. Dotted smartweed is found in moist, low places. Blooms from July to October.

Polygonaceae
Persicaria punctata (Elliott) Small

Dotted Smartweed

---

7. Leaves without scarious sheathing stipules

8. Leaves awl-shaped and sharp pointed

Bushy, much-branched annual, 1 to 4 feet high, forming rounded clumps. Leaves linear, somewhat fleshy, with a sharp prickly point. Flowers axillary with a membranous wing, subtended by a prickle-tipped bract. Plant becomes rigid and spiny at maturity breaking off at ground level and blowing across fields as a "tumbleweed." Very common weed of waste places. Blooms July to October.

Chenopodiaceae
Salsola tragus L.

Russian Thistle

---

8. Leaves not awl-shaped and sharp pointed

9. Calyx colored and petal-like in a cup-shaped involucre

10. Calyx white or pinkish

Perennial with whitish tomentose stems. Leaves mostly basal with a long leafless inflorescence, 1 to 3 feet long. Leaves ovate, undulate, green above and white beneath; 1 to 2 inches long. Calyx white or pinkish; 6-lobed. Stamens 9. Several flowers in an involucre. Involucres sessile and axillary. Common plant in dry rocky places. Blooms from August to November.

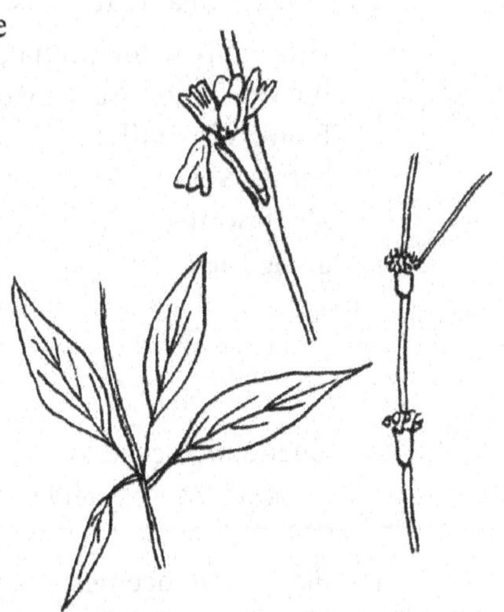

Polygonaceae
Eriogonum elongatum Benth.

Long-Stem Wild Buckwheat

---

10. Calyx yellow

Low perennial with white-woolly leaves. Lower stem woody and covered with old leaves. Leaves ovate, ½ to 1½ inches long. Flowers in a rather dense, terminal cluster. Calyx bright yellow. Flowers from April to July. Found on rocky slopes and canyons in southern Ventura County, particularly in Wildwood Canyon in the Conejo Valley. **RARE.**

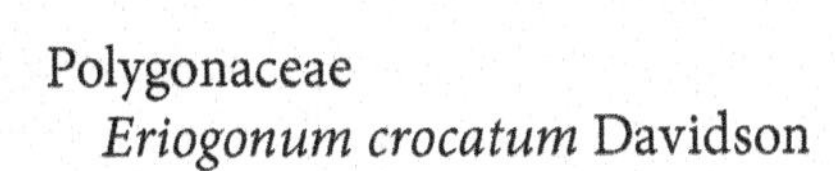
Polygonaceae
Eriogonum crocatum Davidson

Conejo Buckwheat

---

9. Calyx not colored or in a cup-like involucre

10. Leaves entire or wavy margined (Leaves toothed, page 57)

11. Leaves 3-veined from the base. Plant hispid

Low, spreading, gray plant with hispid stems and pubescent leaves. Leaves thick, ovate, palmate veined. Abundant weed along road sides and on dry disturbed open soil. Blooms from May to October.

Euphorbiaceae
Croton setigerus Hook.

Turkey Mullein "Doveweed"

---

11. Leaves with 1 vein from the base

12. Leaves linear

13. Flowers in axillary clusters, appearing woolly

Gray, pubescent annual; prostrate; stems branching from the base. Leaves linear-lanceolate, entire. Flowers in axillary clusters, appearing woolly because of the pubescent calyx. Common weed in alkaline areas. Particularly abundant in salty areas along the coast.

Chenopodiaceae
Bassia hyssopifolia (Pall.) Kuntze

Five-Hook Bassia

---

13. Flowers in racemes; not woolly. Fruit of 4 nutlets, forming an X-shape at maturity

Low, slender, pubescent annual; prostrate or spreading. Leaves linear ¼ to 1½ inches long. Flowers very small, white, less than ⅛ inch long. Fruit becoming 4 nutlets forming an X-shaped figure about ⅛ inch long with a bristly margin. Occasionally found in dry soil. Blooms from March to April. *Pectocarya linearis* has nutlets with the margin bristly the entire length. **Northern Pectocarya,** *Pectocarya penicillata* (Hook. & Arn.) A. DC., nutlets are bristly only at the ends.

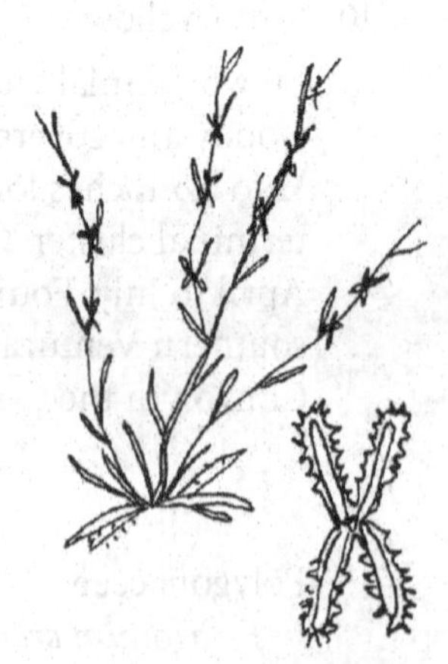

Boraginaceae
Pectocarya linearis (Ruiz & Pav.) DC.
subsp. *ferocula* (I.M. Johnst.) Thorne

Narrow-Toothed Pectocarya

12. Leaves not linear

13. Plant hoary white

Erect perennial with alternate entire leaves. Stem and leaves grayish white, scurfy or hoary. Leaves elliptic to oblong, entire, ¾ to 2 inches long. Inhabitant of dry sandy areas, frequently found on sand at the beach. Blooms from March to October.

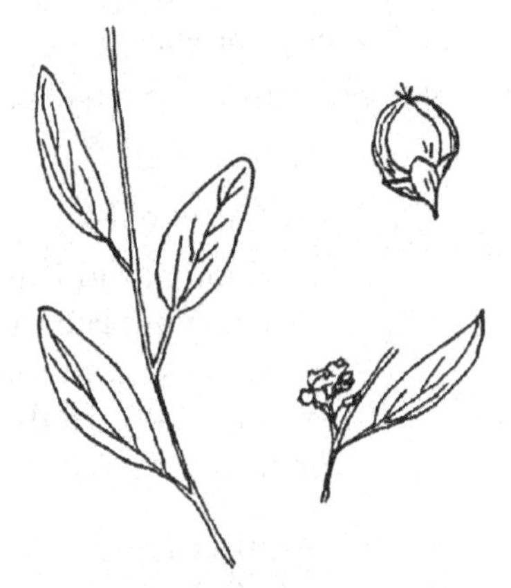

Euphorbiaceae
Croton californicus Müll. Arg.

California Croton

---

13. Plant not hoary white

14. Stems pubescent

15. Plant with strong, unpleasant odor. Inflorescence in axillary clusters

Tall, erect plant, 1 to 3½ feet tall with strong unpleasant odor. Leaves lanceolate; toothed below; subentire or entire above; ¾ to 4 inches long. Flowers in a leafy panicle of spikes. Weed along stream beds. Blooms from June to December.

Chenopodiaceae
Dysphania ambrosioides (L.) Mosyakin & Clemants

Mexican Tea

---

15. Plant not strong odored. Inflorescence in axillary and terminal, more or less prickly clusters

Stout, erect plant, 1 to 4 feet tall, with rough pubescent stems. Leaves glabrous, wavy margined, ovate; 1 to 4 inches long; petioled. Flowers in dense spikes; terminal and lateral. Flowers subtended by prickle-tipped bracts making inflorescence prickly. Common weed of summer. Blooms from June to November. Young leaves may be boiled and eaten. Cook as soon after picking as possible to preserve the flavor.

Amaranthaceae
Amaranthus retroflexus L.

Redroot Pigweed

---

14. Stems glabrous

15. Plant prostrate or spreading (Plant erect, page 57)

16. Plant succulent

17. Fruit a curved capsule. Flowers in a coiled inflorescence

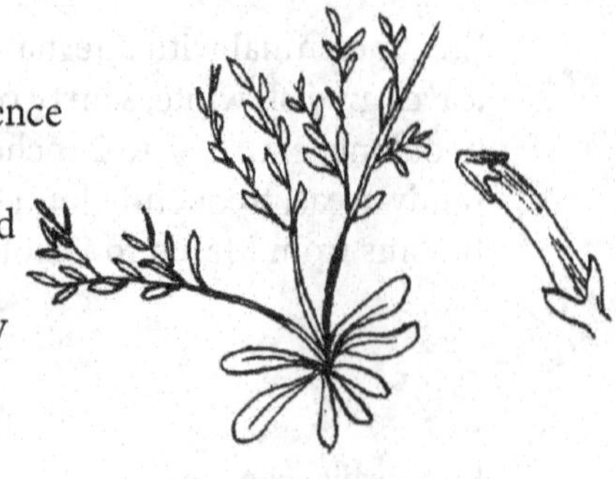

Low annual 2 to 7 inches high, often on the ground. Leaves spathulate, entire, ¾ to 2 inches long; reduced above. Flowers mostly inconspicuous in a coiled inflorescence. Petals 3. Sepals 2. Occasional in sandy or burned over areas. Blooms from March to June.

Montiaceae
Calyptridium monandrum Nutt.

Common Pussypaws

---

17. Fruit a globose capsule. Flowers not in a coiled inflorescence. Common weed

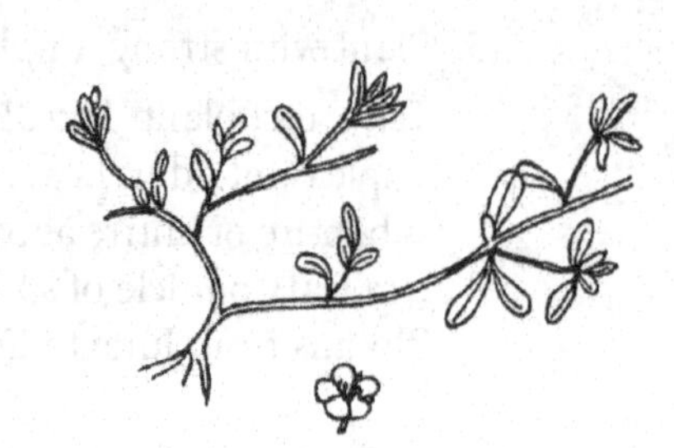

Prostrate annual with fleshy, more or less reddish stems, 4 to 8 inches long. Leaves somewhat thick and fleshy, obovate to spathulate, up to 1 inch long. Petals 5; pale yellow; ⅛ to ¼ inch across. Sepals 2. Stamens 7 to 20. Common weed in waste places. Flowers from May to September. Leaves and young stems of this plant are edible and quite delicious. They may be eaten in a salad or boiled as a potherb.

Portulacaceae
Portulaca oleracea L.

Purslane

---

16. Plant not succulent

17. Flowers in axillary clusters

Glabrous, prostrate plant with stems 4 to 20 inches long. Leaves numerous, crowded; ¼ to 1 inch long. Flowers in axillary clusters subtended by ovate bracts. Occasional weed of waste places. Blooms from July to November. Young leaves are edible.

Amaranthaceae
Amaranthus blitoides S. Watson

Procumbent Pigweed

---

17. Flowers in terminal and axillary spikes

Slender spreading or prostrate plant with glabrous stems. Leaves glabrous, ovate to lanceolate; ¼ to ¾ inch long. Flowers in short, thick axillary and terminal spikes. Bracts present, but inconspicuous and not prickle-tipped. Summer weed of lawn and disturbed ground. Blooms from May to November. Young leaves are edible and should be cooked soon after picking to preserve their flavor.

Amaranthaceae
Amaranthus deflexus L.

Largefruit Amaranth

---

15. Plant erect or ascending

Bushy, much branched, erect plant with glabrous stems. Very similar to the above except for being erect. Leaves spathulate to elliptic, ¼ to 2 inches long; mucro-tipped. Flowers in axillary clusters subtended by prickle-tipped bracts. Plant is easily uprooted by winds in the fall and blown along as a "tumbleweed." Blooms in the summer from June to October.

Amaranthaceae
Amaranthus albus L.

Tumbleweed

---

10. Leaves toothed (at least the lower leaves)

11. Fruits enclosed in two disk-like bracts

12. Perennial with woody base. Plant prostrate or spreading

Prostrate and spreading perennial with branched stems; gray scurfy when young. Leaves numerous, alternate, ¼ to 1 inch long; ovate, dentate or almost entire. Flowers small, axillary. Fruiting bracts becoming red and somewhat swollen at maturity; rhombic; about $^{3}/_{16}$ inch long. Common weed in alkaline soils. Blooms from April to December.

Chenopodiaceae
Atriplex semibaccata R. Br.

Australian Saltbush

---

12. Annual. Plant erect

Erect annual, 1 to 3 feet tall. Leaves grayish scurfy, ovate, sinuate dentate; ¾ to 2 inches long. Upper leaves reduced, entire. Flowers in axillary clusters and terminal spikes. Fruiting bracts rhombic, grayish, flattened with acute apex. Plant of alkaline soils. Blooms from July to October.

Chenopodiaceae
Atriplex rosea L.

Tumbling Orach

---

11. Fruits not enclosed in disk-like bracts

12. Plant with strong unpleasant odor. Herbage glandular pubescent

See *Dysphania ambrosioides*, page 55.

12. Plant not strong odored

13. Flowers in a terminal raceme. Fruit a winged pod, pendant

Glabrous, erect, slender annual 8 to 20 inches tall. Basal leaves dentate, lanceolate, ¾ to 2 inches long. Upper leaves lanceolate, entire, clasping. Flowers minute. Petals white, less than 1/16 inch long. Fruits pendant or hanging downward; flattened, orbicular, with a membranous wing having radiating veins. Fruits 3/16 inch across or less. Inhabitant of brushy or grassy hills and slopes. Blooms from March to May. **Lace-Pod**, *Thysanocarpus laciniatus* Nutt., is similar to *T. curvipes* except that the cauline leaves are not auriculate clasping, and the basal leaves are glabrous and do not occur in a rosette. Blooms March to May.

Brassicaceae
Thysanocarpus curvipes Hook.

Fringe-Pod

---

13. Flowers in a terminal spike or axillary and terminal clusters. Fruits not as above

14. Leaves coarse dentate

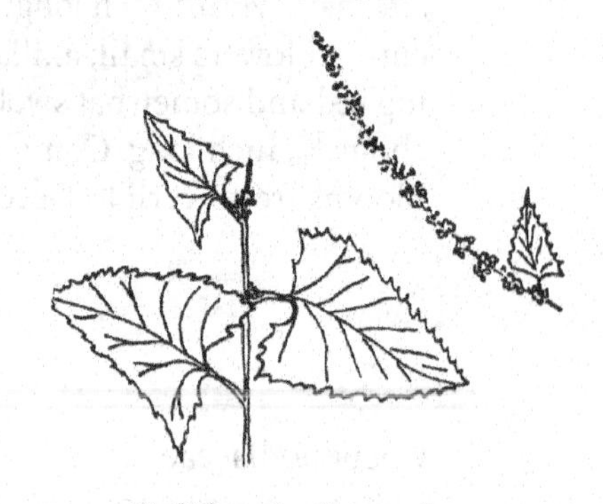

Slightly mealy perennial 1 to 3 feet tall. Leaves alternate, deltoid, coarsely dentate, 1 to 4 inches long. Flowers small, greenish, in dense terminal spikes 12 inches and more long. Root large and carrot-like. Inhabits dry slopes and plains. Blooms from March to June. Roots may be used for soap. Grind the tubers on rocks and mix with water.

Chenopodiaceae
Chenopodium californicum (S. Watson) S. Watson

California Goosefoot

---

14. Leaves shallow toothed to subentire

15. Leaves pale-green, mealy. Plant tall; up to 6 feet tall

Tall, pale-green annual up to 6 feet tall, often forming tall rounded clumps. Herbage mealy and stems red-veined. Leaves glaucous, ovate to lanceolate, shallow toothed or lobed; densely mealy beneath, at least when young. Flowers in dense axillary and terminal clusters. Common weed of summer. Blooms from June to October.

Chenopodiaceae
Chenopodium album L.

Lamb's Quarters

15. Leaves dark green, sparingly mealy. Plant shorter; seldom over 20 inches high

Glabrous, somewhat ill-scented annual with red-veined stem and sparsely mealy herbage; 5 to 20 inches high. Leaves dark green, ovate, ¾ to 2¼ inches long; sinuate dentate. Flowers in dense axillary and terminal clusters, sometimes slightly reddish at maturity. Very common weed, particularly abundant in the spring, but blooms from January to December.

Chenopodiaceae
Chenopodium murale L.

Nettleleaf Goosefoot

---

5. Leaves opposite or whorled
6. Leaves whorled
7. Leaves 6 to 8 in a whorl

Weak-stemmed annual with retrorse hispid stems, climbing over other plants and much tangled. Leaves 6 to 8 in a whorl; linear oblanceolate, bristle-tipped. Flowers small; less than ⅛ inch across; whitish, in the upper axils. Petals 4; sepals mostly obsolete. Fruit bristly, $^{3}/_{16}$ inch across. Common plant in shaded areas. Blooms from March to July.

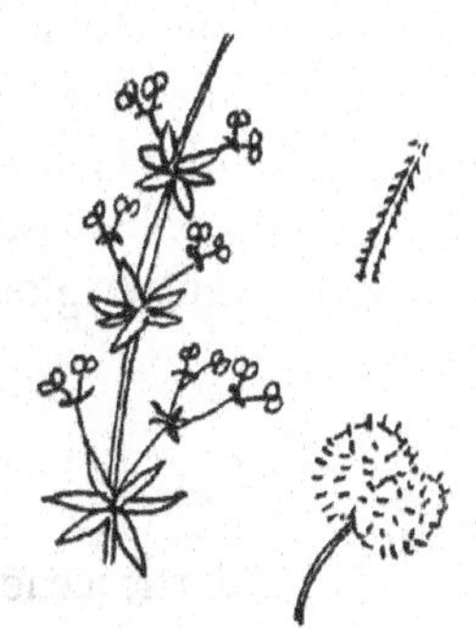

Rubiaceae
Galium aparine L.

Goose Grass

---

7. Leaves 4 in a whorl

Perennial, slightly woody at the base. Stems retrorse hispid, climbing over other plants. Leaves 4 in a whorl, oval to linear oblong, up to ¼ inch long. Flowers yellowish-green, less than $^{1}/_{16}$ inch across. Fruits glabrous, ⅛ inch in diameter. Common plant of chaparral and dry slopes. Blooms from March to June.

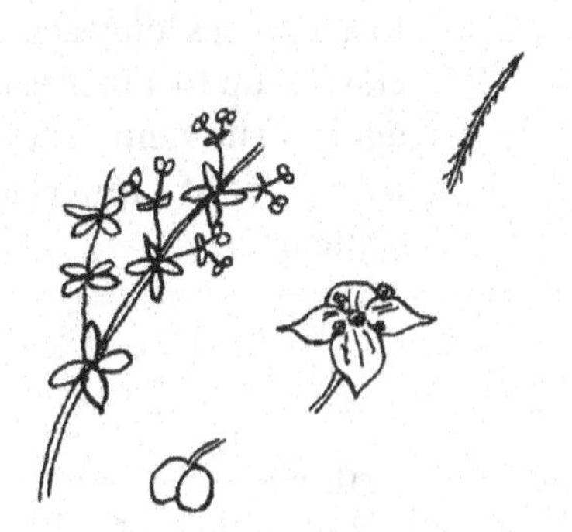

Rubiaceae
Galium nuttallii A. Gray

San Diego or Climbing Bedstraw

---

6. Leaves opposite

7. Plant low, succulent; less than 3 inches high. Leaves minute, ⅛ inch long

Low, glabrous annual, ¾ to 3 inches high with minute, fleshy leaves which may become reddish in age. Leaves opposite, ovate to oblong, ⅛ inch long. Flowers small, axillary. Petals less than $\frac{1}{16}$ inch long. Pygmy-weed is common in masses on dry soil, particularly after a burn. It is tiny and resembles moss from a distance.

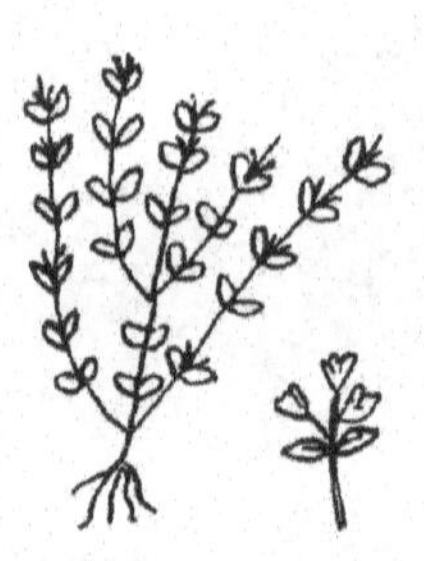

Crassulaceae
Crassula connata (Ruiz & Pav.) A. Berger

Pygmy-Weed

---

7. Plant not as above

8. Plant with stinging hairs. Common along streams and in damp places

9. Leaves less than 1¼ inches long. Plant glabrous except for stinging hairs

Annual with **stinging hairs**. Stems 4 to 20 inches high. Leaves opposite, ovate, serrate, glabrous; ½ to 1¼ inches long. Flowers greenish, in loose clusters ⅜ inch long. Weed of damp places. Blooms from January to April. **Touching or rubbing against the stem or leaf can be very painful**. The leaves are edible, if cooked, but great care must be taken in collecting. Heavy gloves are recommended.

Urticaceae
Urtica urens L.

Dwarf Nettle

---

9. Leaves 2 to 4½ inches long. Herbage pubescent

Perennial with **stinging hairs**. Stems 3 to 8 feet tall. Leaves opposite, petioled, lanceolate-ovate; 2 to 4½ inches long; coarsely serrate. Flowers in loose clusters up to 4 inches long. **Touching or rubbing against the stem or leaf can be very painful**. Common plant of damp places. Blooms from June to September.

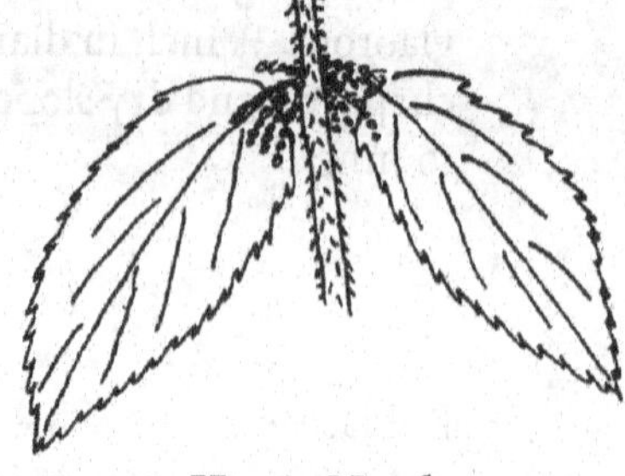

Urticaceae
Urtica dioica L.
subsp. *holosericea* (Nutt.) Thorne

Hoary Nettle

---

8. Plant without stinging hairs

9. Plant with milky juice

10. Plant erect

See *Euphorbia peplus*, page 50.

10. Plant prostrate

11. Stems and fruits with appressed hairs

Prostrate annual with reddish pubescent or hairy stems and milky juice. Leaves opposite, often red-spotted in the center; entire, asymmetrical at base; averaging ⅜ inch long. Flowers minute with narrow white petaloid margin. Ovary appressed hairy. Common weed of waste places. Blooms from May to October. Other taxonomic consideration is *Euphorbia maculata.*

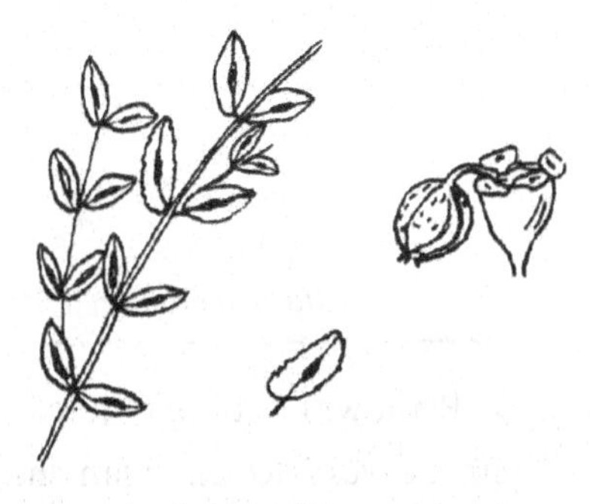

Euphorbiaceae
Chamaesyce maculata (L.) Small

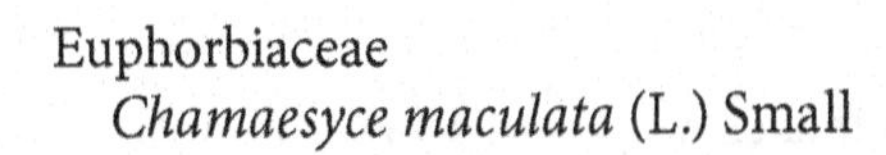

Spotted Spurge

---

11. Stems and fruits glabrous

12. Leaves entire

13. Leaves without thin, white margin

Prostrate perennial with glabrous stems. Leaves round to ovate, opposite, entire; mostly ⅛ inch long or less. Flowers minute with a narrow white margin or appendage and maroon center. Fruits glabrous. Common plant of dry slopes. Blooms most of the year but is most frequently found in the spring. Other taxonomic consideration is *Euphorbia polycarpa.*

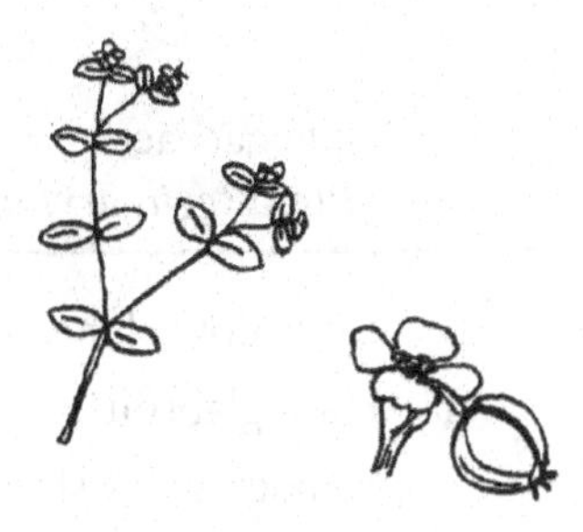

Euphorbiaceae
Chamaesyce polycarpa (Benth.) Millsp.

Smallseed Sandmat

---

13. Leaves with thin, white margin

Prostrate perennial with glabrous stems 2 to 10 inches long. Leaves are oblong; ⅛ to ⅜ inch long with a thin, white margin. Common on dry slopes and fields. Blooms from April to November. Other taxonomic consideration is *Euphorbia albomarginata.*

Euphorbiaceae
Chamaesyce albomarginata
(Torr. & A. Gray) Small

Rattlesnake Weed

---

12. Leaves with a toothed margin

Prostrate annual with glabrous stems 2 to 14 inches long. Leaves are 1 to 5½ inches long; finely toothed; oblong to ovate with linear stipules. Glands have a narrow, white appendage. Capsule is glabrous, $^{1}/_{16}$ inch long. Common in dry, disturbed areas. Blooms from August to October. Other taxonomic consideration is *Euphorbia serpyllifolia.*

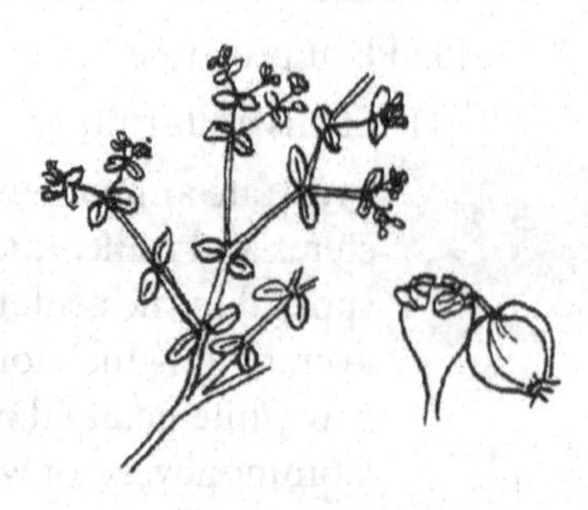

Euphorbiaceae
Chamaesyce serpyllifolia (Pers.) Small

Thyme-Leafed Spurge

---

9. Plant without milky juice

10. Leaves 2-lobed or fan-shaped

Prostrate annual with pubescent stems 4 to 6 inches long. Leaves are opposite, 2-lobed or fan-shaped; somewhat reddish; ¼ to ¾ inch long. Flowers are minute, reddish, with reddish bracts subtending the flowers. Common in shady places. Blooms from March to June.

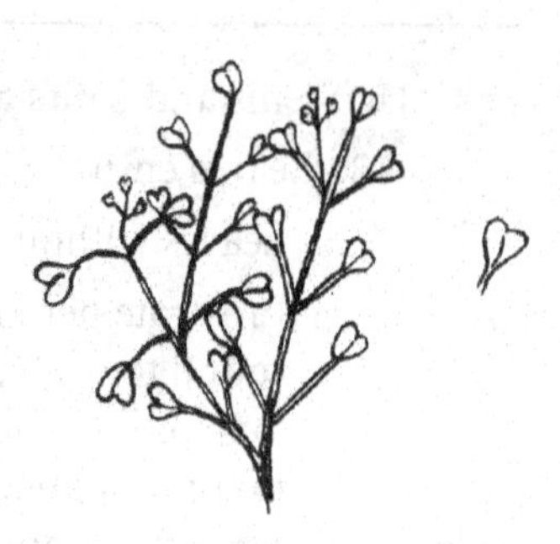

Polygonaceae
Pterostegia drymarioides Fisch. & C.A. Mey.

Woodland Threadstem

---

10. Leaves not 2-lobed or fan-shaped

11. Plant glabrous

Slender, weak-stemmed annual with trailing stems. Leaves opposite, entire, ovate; mostly ¼ to 1 inch long. Flowers small, solitary in leaf axils. Sepals 5; $^{3}/_{16}$ inch long. Petals white; shorter than the sepals, bifid at apex; 4 to 5 in number. Common weed in moist shaded areas. Blooms from February to September. The leaves of chickweed are edible and may be boiled like spinach. The young ones are best because old leaves tend to become stringy.

Caryophyllaceae
Stellaria media (L.) Vill.

Common Chickweed

---

11. Plant sticky, pubescent

Viscid pubescent annual, 4 to 12 inches high. Leaves are opposite, narrow ovate, ⅜ to 1 inch long with obtuse margins. Flowers have 5 white, 2-cleft petals, the same length or shorter than the sepals. Sepals 5; $^{3}/_{16}$ inch long with scarious margins. Fruit a slightly curved capsule, ¼ to ⅜ inch long. Common plant of waste places. Blooms from February to May.

Caryophyllaceae
Cerastium glomeratum Thuill.

Sticky Mouse-Ear Chickweed

---

1. Plants with petals
2. Plants with milky juice. Flowers with petals reflexed and a hood-like crown between the petals and stamens. **Milkweed**
3. Leaves linear, in whorls. Flowers greenish-white. Herbage glabrous

Erect, glabrous plant 1 to 3 feet tall with milky juice. Leaves linear or linear lanceolate, 1½ to 5 inches long; whorled. Flowers in umbels; pedicels ¼ to ⅝ inch long. Corolla greenish-white or lavender tinted; 5-lobed; lobes reflexed. Heads white, incurved, with an incurved tooth-like projection called a horn. Stamens 5. Fruit a pod which opens on one side at maturity. Pod smooth, narrow, sharp-pointed, 2½ to 3½ inches long. Seeds with a tuft of fine white hairs. Common plant in dry open places. Blooms from June to September.

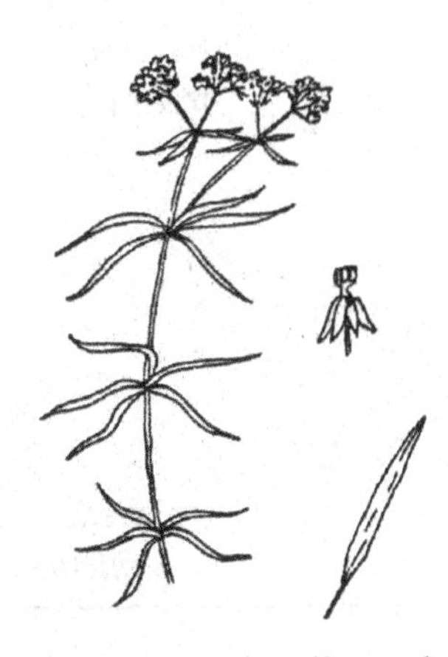

Apocynaceae
Asclepias fascicularis Decne.

Narrow-Leaf Milkweed

---

3. Leaves ovate, opposite. Flowers maroon. Herbage white-woolly

Erect, stout, herbaceous plant, ½ to 2 feet high, with white-woolly herbage and milky juice. Leaves opposite, ovate, 2 to 6 inches long; densely white-woolly. Flowers in umbels. Calyx white-woolly. Corolla 5-lobed; lobes purple; reflexed at maturity. Hoods maroon, without horn-like appendages. Fruit an ovoid pod, hoary, opening on one side at maturity; mostly 2 to 3 inches long. Found on dry chaparral slopes, especially near the coast. Not common. Blooms from April to July.

Apocynaceae
Asclepias californica Greene

California Milkweed

---

2. Plants without milky juice. Flowers not having reflexed petals with accessory hoods

3. Flowers in simple or compound umbels. Ovary inferior.

Leaves compound **Apiaceae—Parsley/Carrot/Celery Family**

(Flowers, not as above, page 68)

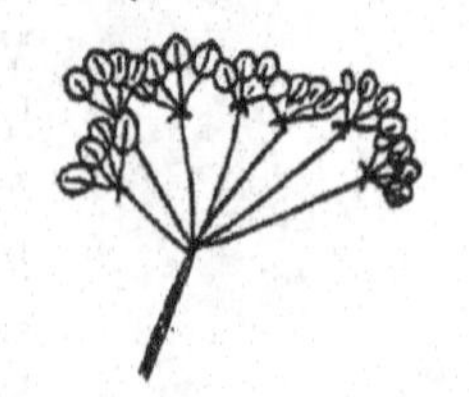

Umbel Inflorescence

---

4. Flowers yellow or red (Flowers white, page 66)

5. Umbels capitate and ball-like. Leaves palmate compound

6. Flowers yellow

Erect glabrous plant, ½ to 1½ feet high. Leaves mostly basal; palmately 3 to 5 parted, coarsely serrate with spinose teeth. Petioles winged. Flowers yellow, in a compact umbellate cluster, about ¼ inch across. Fruits ovoid, ¼ inch long, with hooked bristles. Inhabitant of grassy slopes of the chaparral. Blooms in the spring from March to April.

Apiaceae

Sanicula arguta J.M. Coult. & Rose

Snakeroot

---

6. Flowers purple-red

Perennial 6 to 18 inches high with basal leaves. Leaves are 2 to 5 inches long and 3 to 7 parted. Divisions are further lobed or cleft. Flowers are purple-red in a compact umbel. Fruits are ovoid, ⅛ to ¼ inch long with prickles. Found on open slopes. Blooms from March to May.

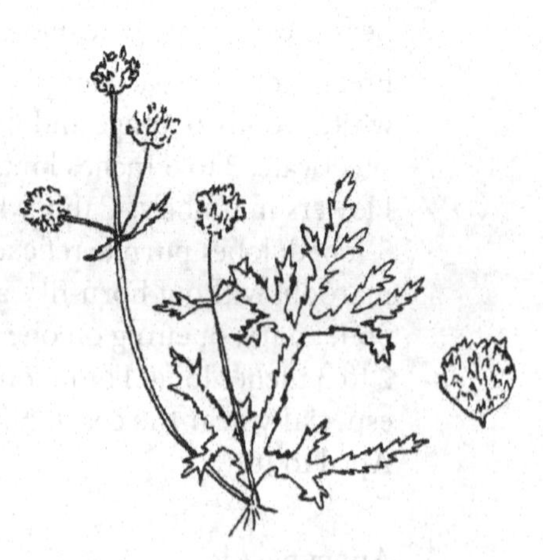

Apiaceae

Sanicula bipinnatifida Hook.

Purple Sanicle

---

5. Umbels more open; not ball-like. Leaves pinnate compound

6. Plant more than 3 feet tall. Herbage with a pronounced licorice odor

Tall perennial, 3 to 6 feet high; glabrous, with a pronounced odor of licorice (anise). Leaves pinnately compound into very fine (filiform) divisions. Petioles sheathing. Flowers yellow, in large compound umbels. Fruits oblong, about ¾ inch long with acute ribs. Very common summer plant of field and waste places. Blooms from May to September. Leaf petioles may be eaten raw or cooked.

Apiaceae
Foeniculum vulgare Mill.

Fennel

---

6. Plant less than 3 feet tall. Herbage without licorice odor

7. Leaves dissected into many linear divisions

Plant ½ to 1½ feet high, sparsely pubescent with a purplish stem below. Leaves pinnately dissected into many linear divisions. Flowers yellow. Fruits ovate, ¼ to ⅜ inch long, with a broad membranous wing. Body of fruit with longitudinal ribs; glabrous at maturity. Relatively common spring plant in the chaparral hills. Blooms from February to May. Roots may be eaten raw and have a taste of celery.

Apiaceae
Lomatium utriculatum (Torr. & A. Gray) J.M. Coult. & Rose

Common Lomatium

---

7. Leaf divisions, oblong or ovate

8. Fruits, roundish with broad wings

Glabrous, perennial plant ¾ to 2 feet high. Leaves arising from lower portion of stem, long petioled, pinnate compound into broad ovate or cuneate divisions. Leaflets dentate and often lobed. Flowers in compound umbels, long peduncled. Petals yellow. Fruits flattened, with broad wings. Main body of fruits longitudinally ribbed. Fruits ¼ to ⅝ inch long.

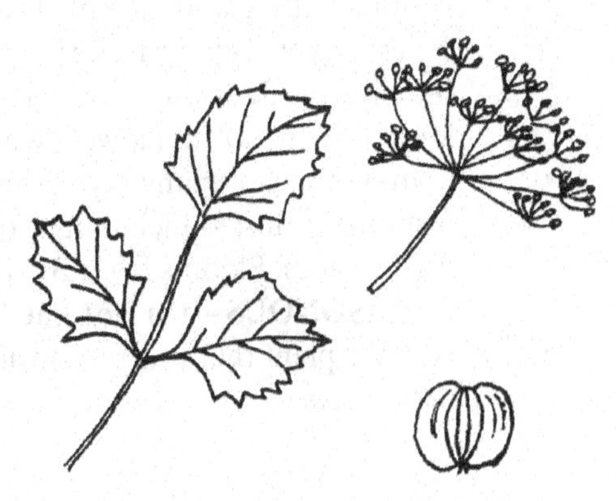

Apiaceae
Lomatium lucidum (Torr. & A. Gray) Jeps.

Shiny Lomatium

---

8. Fruits, oblong without wings

Glabrous perennial with ribbed stem, 1 to 2½ feet high. Leaves are basal; pinnate compound with ovate leaflets, 1 to 3 inches long and sharply toothed. Flowers are yellow. Fruits are oblong, ribbed, ¼ to ⅝ inch long. Occurs on slopes from Santa Barbara to San Diego. Blooms from April to June.

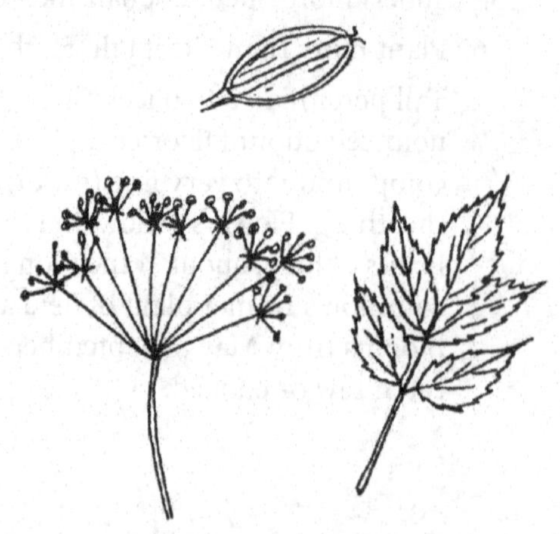

Apiaceae
Tauschia arguta (Torr. & A. Gray) J.F. Macbr.

Southern Tauschia

---

4. Flowers white
   5. Leaves finely dissected into narrow or linear divisions
      6. Fruits without wings
         7. Fruit heart-shaped; ribs inconspicuous. Stem not purple dotted

Glabrous annual, 2 to 20 inches high. Leaves finely dissected into filiform divisions. Flowers white in umbels. Fruit ellipsoid, up to ½ inch long, minutely roughened, without wings or longitudinal ribs. Occurs in dry, sandy valleys and hills. Blooms from March to April. Used by the Native Americans to supplement their diet.

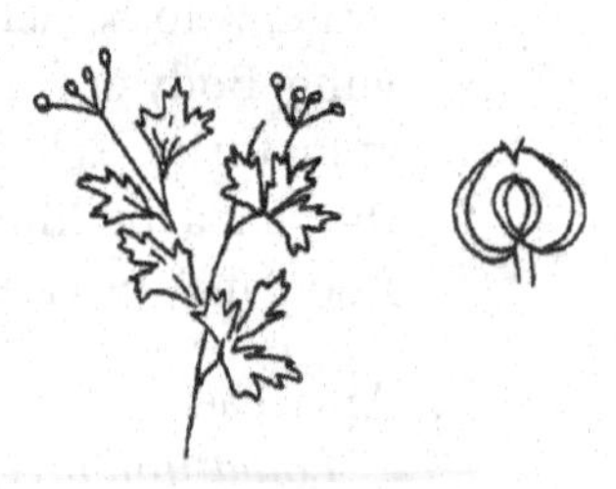

Apiaceae
Apiastrum angustifolium Nutt.

Wild Celery

---

7. Fruit ovoid; ribs prominent. Stem purple dotted

Biennial with glabrous, purple dotted stem; 1½ to 10 feet high. Leaves are pinnate compound into narrow divisions. Upper leaves are sessile; lower ones petioled. Flowers white in a compound umbel of many rays. Fruits are ovoid with prominent ribs; ⅛ inch long. Common in low, waste places. Blooms from April to July. Herbage is **POISONOUS—Do Not Eat.** This is supposedly the plant that was responsible for the death of Socrates.

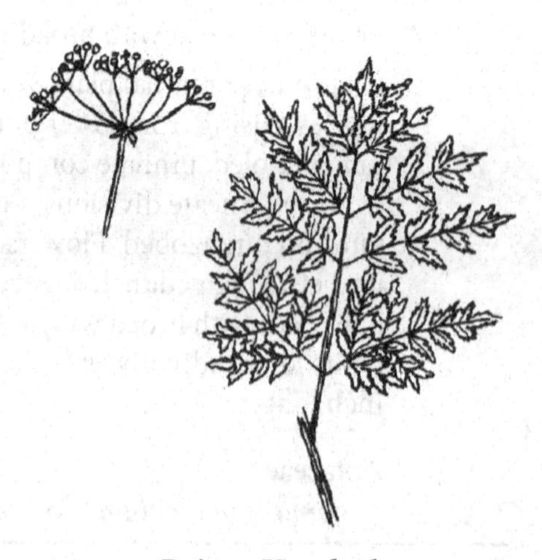

Apiaceae
Conium maculatum L.

Poison Hemlock

---

6. Fruits winged; longitudinal ribs on main body

Low, pubescent plant 4 to 16 inches high; purplish below. Leaves divided into linear divisions; mostly basal. Flowers white in compound umbels. Fruit oblong, mostly ¼ to ½ inch long, pubescent, with broad membranous wings. Ribs apparent on the main body of fruit. Common plant on dry ridges, blooming from March to June.

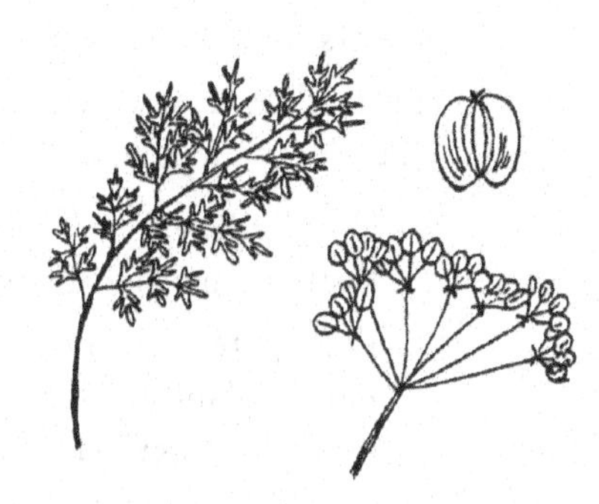

Apiaceae
Lomatium dasycarpum (Torr. & A. Gray) J.M. Coult. & Rose

Woolly Lomatium

---

5. Leaves pinnatifid into broader divisions

6. Plant glabrous with odor and taste of celery. Fruit ovoid

Glabrous annual, 1½ to 4 feet high with odor and taste of celery. Leaves basal, pinnate compound into 5 to 9 lobed or toothed leaflets. Flowers white, in compound umbels. Umbels sessile or short pedicelled. Fruits ovoid; glabrous with prominent ribs. Common plant in wet places. Blooms from May to July. This is common celery which has escaped cultivation. It is edible and delicious. The upper leaves and stems can accumulate high levels of nitrate, so discard them.

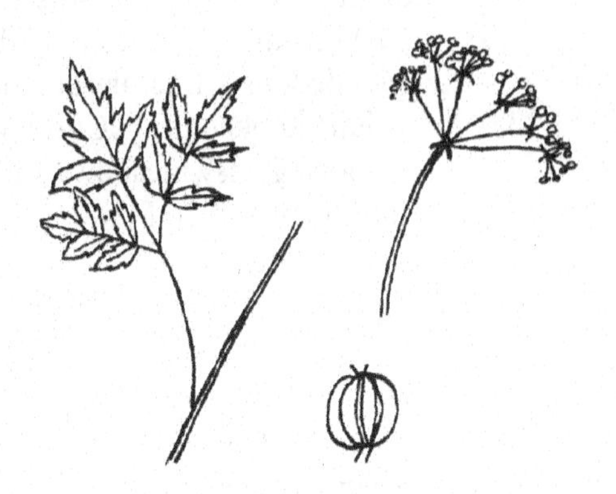

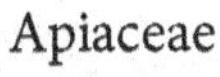

Apiaceae
Apium graveolens L.

Celery

---

6. Plant villous pubescent; without odor or taste of celery. Fruits oblong; hispid

Villous, pubescent perennial 1 to 3 feet high. Leaves compound into ovate leaflets, coarsely, serrate or lobed. Flowers white or greenish-yellow in compound umbels. Fruits ½ or ¾ inch long, oblong, with bristly ribs; narrowed above and below. Fruits hispid and without wings. Occasionally found in shaded areas. Blooms from March to June. Roots of this species have a strong anise flavor and can be used for seasoning.

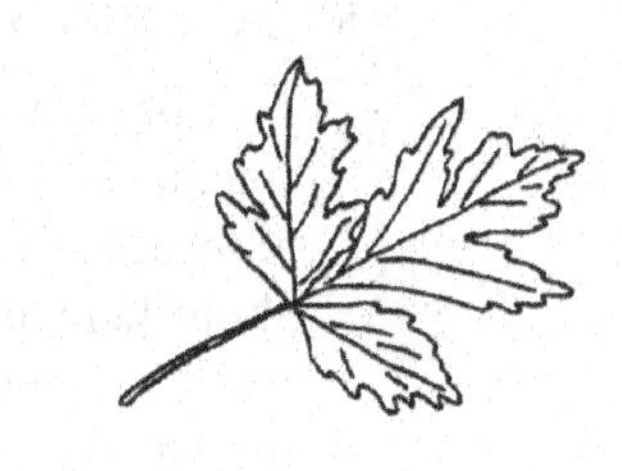

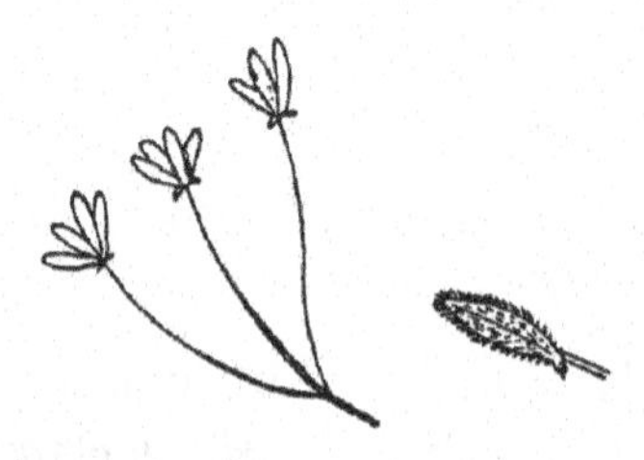

Apiaceae
Osmorhiza brachypoda Torr.

Sweet-Cicely

---

3. Flowers not as above, in all respects

4. Flowers regular (Flowers, irregular, page 112)

5. Petals distinct and separate; not fused. Calyx may or may not be fused (Petals fused, page 101)

6. Petals more than 6 (Petals 6 or less, page 69)

7. Pistils numerous. Flowers yellow. Plants not succulent

Erect perennial, 1 to 2 feet high. Stems mostly glabrous, branched above. Basal leaves long petiolate; blades palmately 3 to 5 lobed or parted; the lobes again lobed into finger-like parts. Petals, leaves, and petioles appressed pubescent to hirsute. Flowers terminal, bright yellow. Petals mostly 9 to 16 in number. Sepals pale-green, reflexed. Stamens numerous. Pistils several (5 to 35) becoming achenes in fruit. Achenes rounded, disk-like, about ⅛ inch long with a short hooked beak. Common spring plant of moist grassy meadows and slopes. Blooms from February to May.

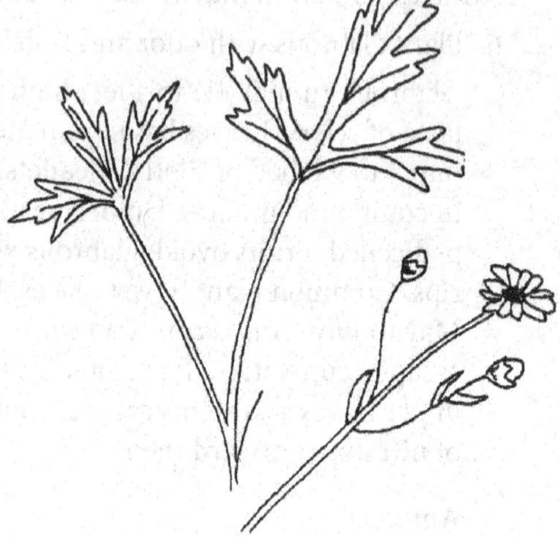

Ranunculaceae
Ranunculus californicus Benth.

California Buttercup

---

7. Pistil 1. Plant succulent; of coast and salt marsh

8. Leaves linear; ¼ to ¾ inch long

Low branched annual, mostly prostrate. Leaves linear, ¼ to ¾ inch long. Flowers solitary in the axils of leaves. Petals numerous, linear, white, about ⅛ inch long. Sepals 5; often reddish in the bud. Stamens numerous. Common plant of salt marshes along the coast. Blooms from April to November.

Aizoaceae—Iceplant Family

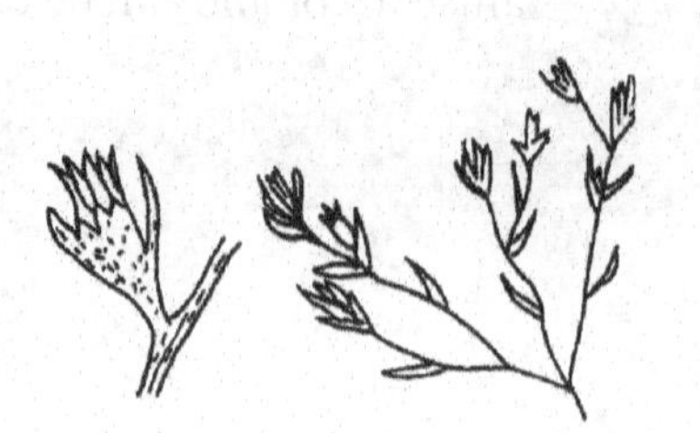

Aizoaceae
Mesembryanthemum nodiflorum L.

Slender-Leaved Iceplant

---

8. Leaves wider

9. Herbage with large crystalline (ice-like) vesicles

Very succulent, prostrate annual with numerous crystalline vesicles on stem and leaves. Leaves ovate, succulent. Flowers white to reddish. Petals numerous, about ¼ inch long. Stamens numerous. Common in salt marshes along the coast. This plant is well named because of the numerous ice-like crystals on the herbage. Blooms from March to October.

Aizoaceae
Mesembryanthemum crystallinum L.

Crystalline Iceplant

---

9. Herbage without prominent ice-like crystals

10. Petals rose-violet

Succulent perennial forming extensive mats over dunes and other areas. Leaves thick, 3-sided, 2½ to 4 inches long. Flowers rose-violet with numerous petals and stamens; 1 to 2 inches across. Petals linear. Common on dunes and bluffs along the coast. Blooms from April to September.

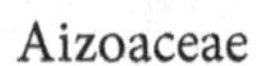

Aizoaceae
Carpobrotus chilensis (Molina) N.E. Br.

Sea Fig

---

10. Petals yellow

Succulent perennial similar to sea fig. Leaves thick, 3-angled, slightly curved, with a serrate edge on the lower angle. Flowers yellow. This species has been extensively planted on dunes and along the highways along the coast.

Aizoaceae
Carpobrotus edulis (L.) N.E. Br.

Freeway Iceplant

---

6. Petals 6 or less

7. Petals dark red to brown

Spring annual, 1 to 2 feet high, glabrous, quickly wilting when picked. Leaves palmately divided into finger-like divisions. Flowers large, solitary, terminal. Petals red-brown, ½ to 1 inch long, almost as wide. Sepals green, 5 or 6. Stamens many. Pistils 2 to 5. Flowers often subtended by reduced leaves. Common early spring plant of the chaparral; usually in woodsy, shaded areas. Blooms from January to March.

Paeoniaceae
Paeonia californica Nutt.

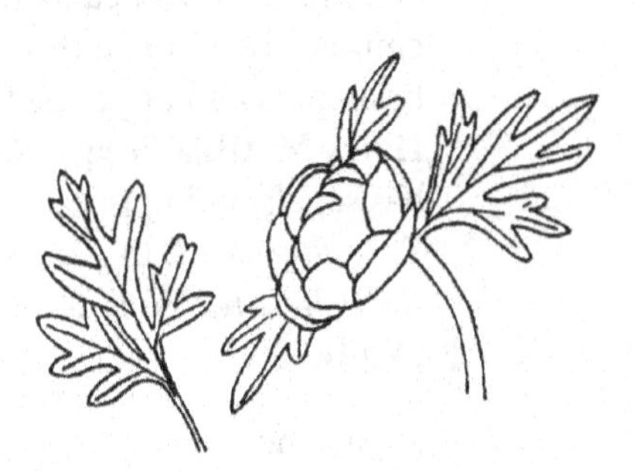

California Peony

---

7. Petals not dark red to brown

8. Petals 6

9. Leaves opposite. Low annual

10. Plant villous. Leaves linear, entire

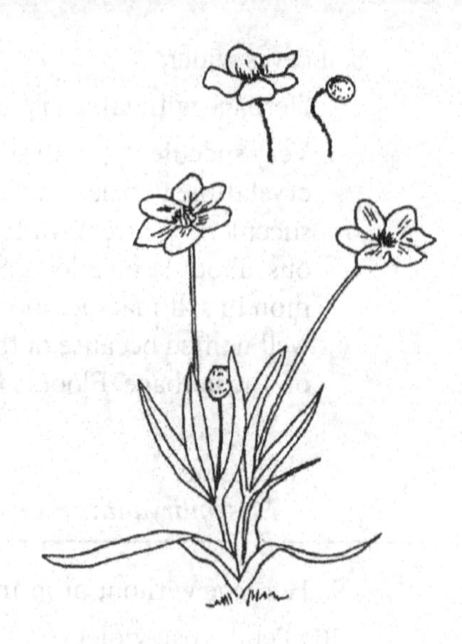

Low annual with many stems from base. Stems with numerous, soft, spreading hairs. Leaves opposite, entire, ¾ to 2 inches long. Flowers solitary, terminal, on long almost leafless stems 4 to 8 inches long. Sepals 3; with long hairs, making buds hairy. Petals 6; cream-colored; ¼ to ⅝ inch long. Stamens numerous. Pistil 1. Occasionally abundant, especially on dry sandy places or after burns. Blooms from March to May.

Papaveraceae
Platystemon californicus Benth.

Cream Cups

---

10. Plant glabrous. Leaves ovate-spathulate toothed

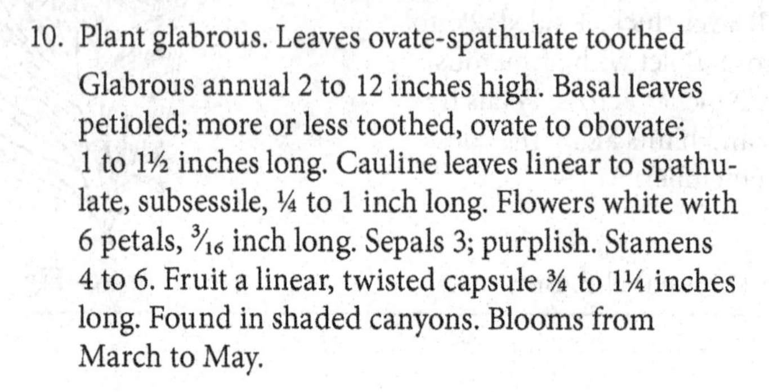

Glabrous annual 2 to 12 inches high. Basal leaves petioled; more or less toothed, ovate to obovate; 1 to 1½ inches long. Cauline leaves linear to spathulate, subsessile, ¼ to 1 inch long. Flowers white with 6 petals, 3/16 inch long. Sepals 3; purplish. Stamens 4 to 6. Fruit a linear, twisted capsule ¾ to 1¼ inches long. Found in shaded canyons. Blooms from March to May.

Papaveraceae
Meconella denticulata Greene

Small-flowered Meconella

---

9. Leaves alternate. Tall perennial, woody at the base

10. Flowers large, white. Petals 2 to 4 inches long

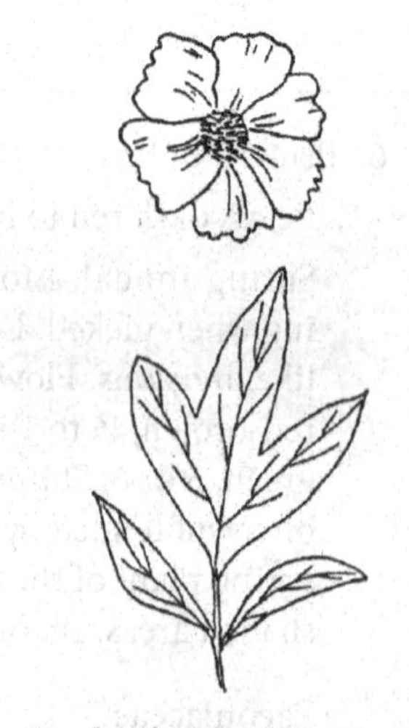

Glabrous perennial, 3 to 7 feet tall, woody at the base. Leaves gray-green, alternate, pinnatifid. Flowers large, showy, at ends of branches. Petals white, 1½ to 4 inches long. Sepals 3; stamens many; pistil 1. Occasionally found in chaparral areas. Frequently cultivated for its beautiful large flowers. Two species commonly occur in the chaparral. *Romneya coulteri* has sepals and upper peduncle glabrous, while the **Hairy Matilija Poppy**, *Romneya trichocalyx* Eastw., has sepals and upper peduncle bristly. Petals of *R. coulteri* are 2 to 4 inches long; those of *R. trichocalyx* 1½ to 3 inches. Blooms from May to July.

Papaveraceae
Romneya coulteri Harv.

Coulter's Matilija Poppy

---

10. Flowers small, purple. Petals ¼ inch long

Erect plant, woody at the base, 20 to 72 inches high. Stems glabrous, pale green. Leaves linear to linear oblong, ⅜ to 1¼ inches long, alternate, entire. Flowers with a cylindric floral tube ¼ inch long and 6 purple petals. Flowers solitary in the axils. *Lythrum* grows in moist places and blooms from April to October.

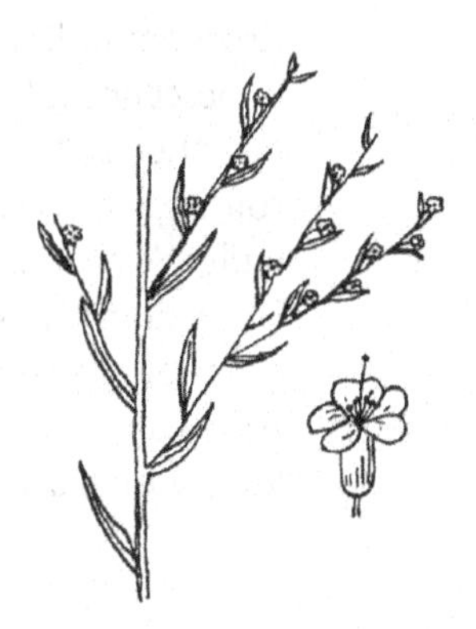

Lythraceae
Lythrum californicum Torr. & A. Gray

California Loosestrife

---

8. Petals less than 6
9. Petals 5 (Petals less than 5, page 87)
10. Stamens fused into a column around the styles

Malvaceae—Mallow Family

11. Stems prostrate. Petals yellowish. Plant of salt marsh

Prostrate perennial with whitish stellate pubescence. Leaves round reniform, dentate, mostly ½ to 1½ inches wide. Petals 5, yellowish, ½ inch long. Inhabitant of saline places such as coastal salt marshes. Blooms from May to October.

Malvaceae
Malvella leprosa (Ortega) Krapov.

Alkali-Mallow

---

11. Stems erect. Petals not yellow
12. Flowers showy, in a few-flowered, terminal raceme. Petals ½ to 1 inch long

Erect perennial with sparsely pubescent stems. Leaves palmately divided or parted and then lobed again; long petiolate and mostly but not entirely basal. Flowers showy, in terminal racemes. Petals ½ to 1 inch long, lavender-pink; usually veined with white. Inhabitant of the chaparral. Blooms in the spring from March to June.

Malvaceae
Sidalcea malviflora (DC.) A. Gray

Checkerbloom

---

12. Flowers inconspicuous in leaf axils. Petals less than ½ inch long

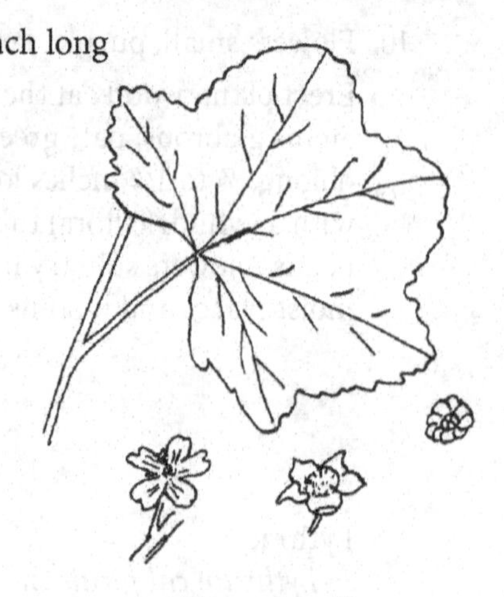

Erect annual, mostly 1 to 3 feet tall with sparse pubescence. Flowers hidden in the leaf axils and often unnoticed. Petals white to pink, ¼ inch long. Leaves roundish or heart-shaped; crenate, ¾ to 3 inches wide. Very common weed. Blooms most of the year. Particularly abundant in February and March. Also occasionally found is **Bull Mallow**, *Malva nicaeensis* All., which differs in having longer petals, ⅜ to ½ inch long, which are 1 to 2 times the length of the calyx.

Malvaceae
Malva parviflora L.

Cheeseweed

---

10. Stamens not fused into a column around the styles
11. Leaves compound or pinnatifid dissected (Leaves simple, page 75)
12. Stamens more than 10 (Stamens not more than 10, page 73)
13. Flowers yellow. Leaves silvery beneath

Glabrous perennial with runners rooting at the nodes. Leaves pinnate compound with 7 to 31 leaflets. Leaflets ovate, sharply serrate, 1 to 2 inches long, green above, silvery beneath. Flowers solitary, terminal, on a scape. Petals 5; yellow; ⅜ inch long. Sepals 5, ¼ inch long. Stamens 20 to 25. Pistils many. Occurs around edges of salt marshes. Blooms from April to September.

Rosaceae
Potentilla anserina L.
subsp. *pacifica* (Howell) Rousi

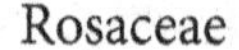

Pacific Silverwood

---

13. Flowers creamy white. Leaves light green beneath

Erect perennial herb, 1 to 3 feet tall, with sticky villous herbage. Leaves pinnate compound; 5 to 9 leaflets. Leaflets obovate, serrate, ½ to 1½ inches long; green above, lighter beneath. Flowers in terminal clusters. Petals 5; pale yellow to creamy white; about ¼ inch long. Sepals 5; same length as petals. Stamens 25. Pistils many. Common inhabitant of the coastal mountains. Blooms from May to July.

Rosaceae
Drymocallis glandulosa (Lindl.) Rydb.

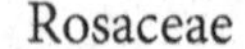

Sticky Cinquefoil

---

12. Stamens not more than 10

13. Flowers yellow

14. Leaves 3-foliate. Leaflets heart-shaped

15. Petals ½ to 1 inch long

Glabrous perennial with scaly bulbs and basal leaves. Petioles 4 to 8 inches long. Leaves 3-foliate. Leaflets obcordate-bilobed, up to 1 inch long. Flowers yellow, showy, ½ to 1 inch long. Petals, sepals, and styles 5; stamens 10, 5 long and 5 short. Occurs in orchards and fields. It is rapidly becoming naturalized. Blooms November to March.

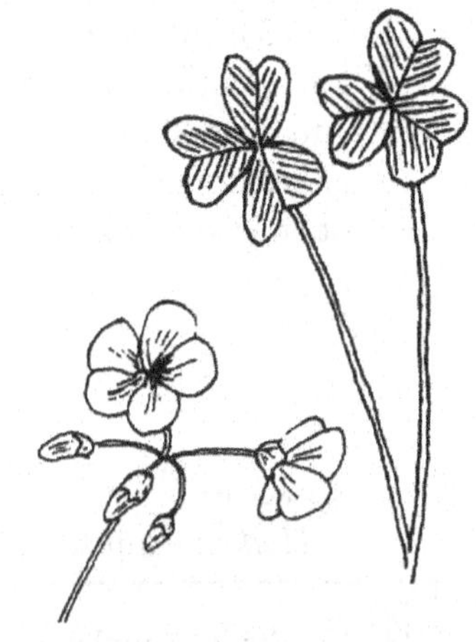

Oxalidaceae
Oxalis pes-caprae L.

Bermuda Buttercup

---

15. Petals less than ½ inch long

Low, slightly pubescent plant with creeping stems. Leaves not all basal, but otherwise similar to the above. Flowers yellow, smaller than the above, petals averaging ¼ inch long. Common weed in lawn and garden. Blooms most of the year.

Oxalidaceae
Oxalis corniculata L.

Creeping Wood-Sorrel

---

14. Leaves 8 to 14 foliate. Common weed. Fruit a bur

Prostrate, spreading annual. Leaves pinnate compound with 4 to 7 pairs of leaflets. Leaflets elliptic. Flowers yellow, solitary, in the axils. Petals 5; about ⅛ inch long. Sepals 5; stamens 10. Fruit becoming a bur with 2 to 4 sharp, spreading spines. A troublesome weed of waste places. Blooms during the summer, from April to October.

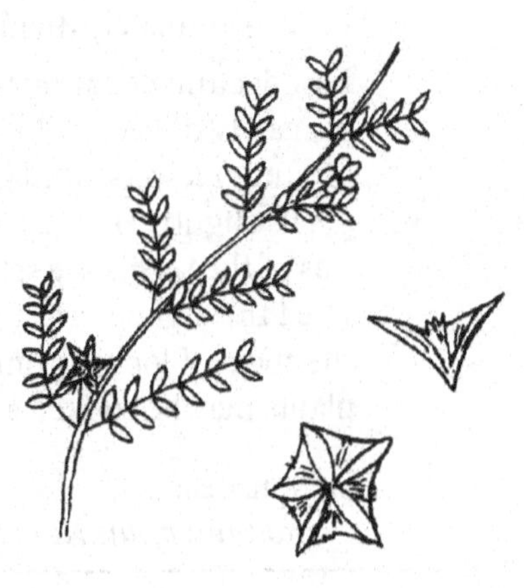

Zygophyllaceae
Tribulus terrestris L.

Puncture Vine

---

13. Flowers not yellow

14. Flowers white

Erect perennial 8 to 20 inches tall with glabrous to glandular, hairy herbage. Leaves pinnate compound into 5 to 10 pairs of leaflets. Leaflets ⅜ to 1¼ inches long, toothed. Flowers white with 5 petals, 10 stamens, and 40 to 80 pistils. Sepals 5, 3/16 inch to ¼ inch long, alternating with shorter bractlets. *Horkelia* grows in open woods. Blooms from April to September.

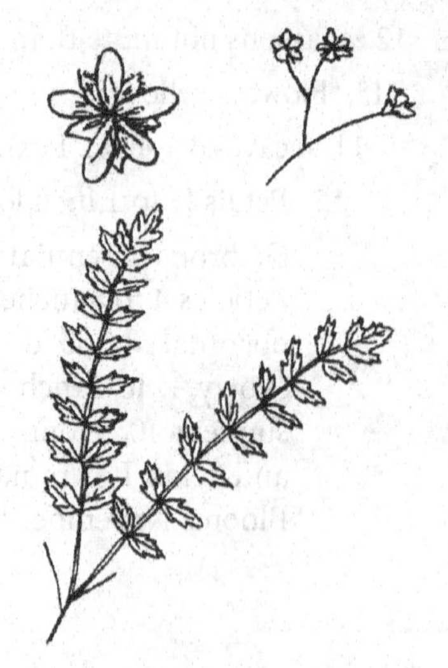

Rosaceae
Horkelia cuneata Lindl.

Wedge-Leaf Horkelia

---

14. Flowers red-violet. Fruit long beaked

15. Beak of fruit ¾ to 1½ inches at maturity

16. Leaflets broad, coarsely toothed

Low, somewhat fleshy annual, decumbent or erect. Leaves pinnate compound with scarious stipules. Leaflets serrate or lobed. Flowers in a terminal umbel. Peduncles 2½ to 8 inches long. Pedicels ¼ to ¾ inch long. Petals 5; red-violet; 5/16 inch long. Sepals 5; stamens 5. Fruit long beaked (elongated style), separating and coiling at maturity. Common weed throughout area; in fields, lawns, and waste places. Blooms from February to May.

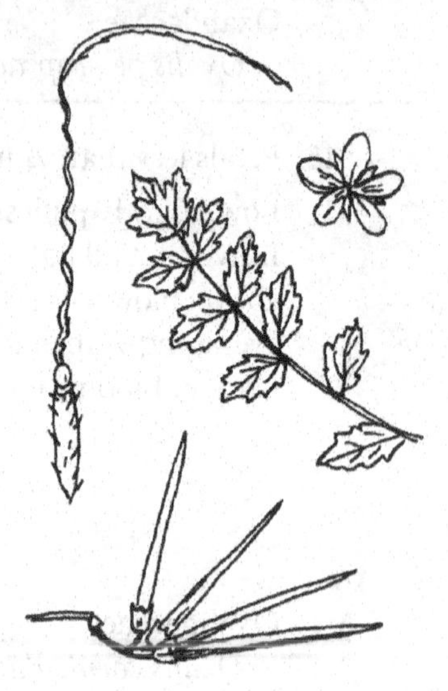

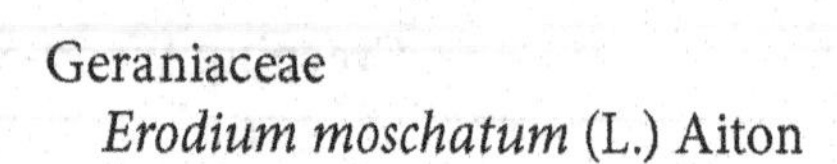

Geraniaceae
Erodium moschatum (L.) Aiton

Greenstem Filaree

---

16. Leaflets pinnately divided into narrow divisions

Low decumbent annual with slender stems. Leaves pinnately dissected, first forming a rosette on the ground. Flowers similar to the above but with the petals slightly smaller, averaging ¼ inch long. The tips of the sepals are setose or hairy. A common weed throughout area. Usually more abundant than the above. Blooms from February to May. Young plants may be cooked as a potherb or used in a salad.

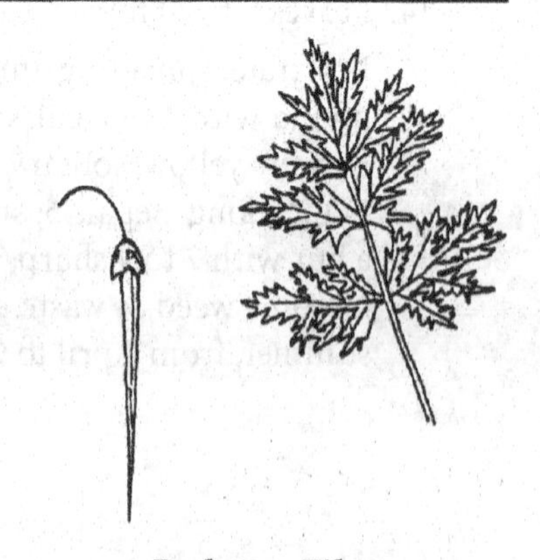

Geraniaceae
Erodium cicutarium (L.) Aiton

Redstem Filaree

---

15. Beak of fruit 3½ to 5 inches long at maturity

Annual with prostrate or erect stems, retrorsely hispid. Leaves pinnatifid dissected with scarious stipules. Flowers in terminal umbels as above. Petals 5; lavender; about ⅝ inch long. Beaks long; 3½ to 5 inches at maturity. Inhabitant of grassy areas and roadsides, but less common than the above.

Geraniaceae
Erodium botrys (Cav.) Bertol.

Long-Beaked Storksbill

---

11. Leaves simple; sometimes deeply lobed

12. Leaves lobed (Leaves toothed or entire, page 77)

13. Leaves palmately lobed; round-cordate to reniform

14. Stamens 10

Perennial herb with pubescent stems, 10 to 20 inches high. Leaves mostly basal. Basal leaves roundish and shallowly lobed. Cauline leaves palmately 3-cleft. Flowers white, in terminal racemes. Petals 5; ¼ to ⅜ inch long; entire or 3-lobed at apex. Sepals 5; stamens 10. Occurs in shaded areas of coastal chaparral. Blooms March to June.

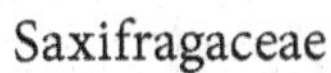

Saxifragaceae
Lithophragma heterophyllum (Hook. & Arn.) Torr. & A. Gray

Hillside Woodland Star

---

14. Stamens 5

15. Leaves shallow lobed. Petals little longer than calyx

Stout perennial, 12 to 32 inches high with glandular villous stems. Leaves round-cordate, 2 to 5 inches wide on petioles 3 to 6 inches long. Blades shallow lobed, crenate toothed. Flowers in panicles 1 to 6 inches long. Petals white, not much longer than calyx. Sepals 3/16 inch long. *Boykinia* grows in wet places in canyons and is common along streams. Blooms from June to July.

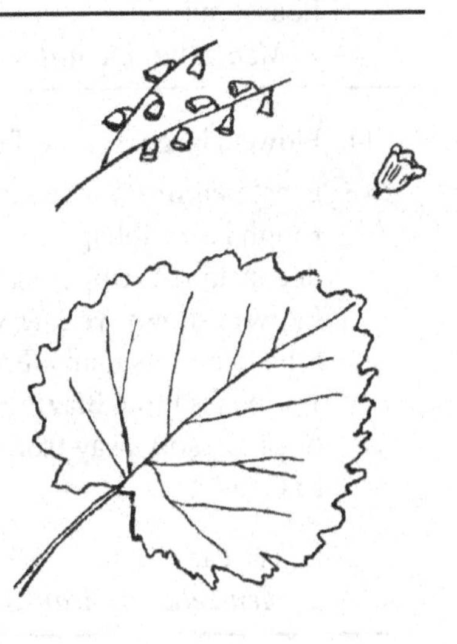

Saxifragaceae
Boykinia rotundifolia A. Gray

Round-Leaved Boykinia

---

15. Leaves deeply lobed. Petals distinctly longer than calyx

Erect perennial 8 to 24 inches high with gland-tipped hairs. Leaves in a basal rosette and also on main stem. Blades reniform to heart-shaped, palmately cleft into 5 to 7 lobes with bristle teeth. Lower leaves ¾ to 3 inches wide. Upper leaves reduced. Flowers white in a many-flowered inflorescence; perigynous. Petals 5, distinctly longer than calyx. Brook foam grows in moist, shaded areas from the Santa Monica Mountains north.

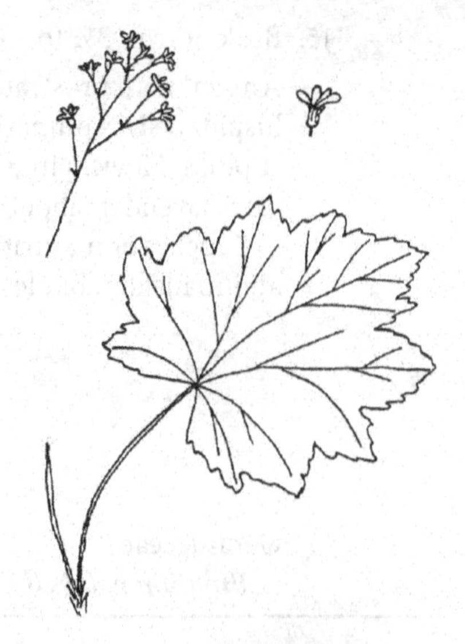

Saxifragaceae
Boykinia occidentalis Torr. & A. Gray

Coast Boykinia

---

13. Leaves pinnately lobed

14. Flowers small, inconspicuous. Petals ⅛ to 3/16 inch long

Erect annual 8 to 20 inches tall with rough grayish stems. Leaves rough hispid with a sandpapery texture enabling them to stick to clothes. Leaves sinuately toothed or lobed, 1 to 3 inches long. Flowers small, in dense clusters at tips of branches, often concealed by green leafy bracts. Petals 5; pale yellow, ⅛ to 3/16 inch long. Sepals 5; stamens many. Common in disturbed places such as burns. Blooms from April to July.

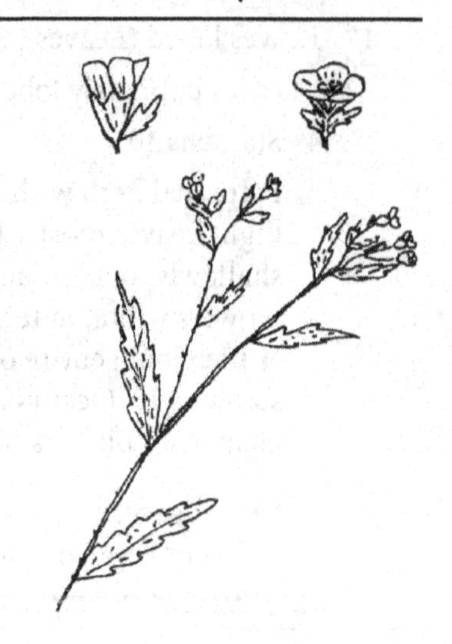

Loasaceae
Mentzelia micrantha (Hook. & Arn.) Torr. & A. Gray

Stickleaf

---

14. Flowers large, showy. Petals 1½ to 3 inches long

Erect perennial with stems 8 to 30 inches high and with rough hairy foliage. Basal leaves lanceolate, shallow to deeply lobed. Upper leaves sinuate lobed or toothed. Flowers showy, yellow, with 5 petals, 1½ to 3 inches long. Stamens many. Fruit a cylindrical capsule ⅝ to 1½ inches long. Blazing star may be found in rocky or sandy soil away from the coast. Blooms from June to October.

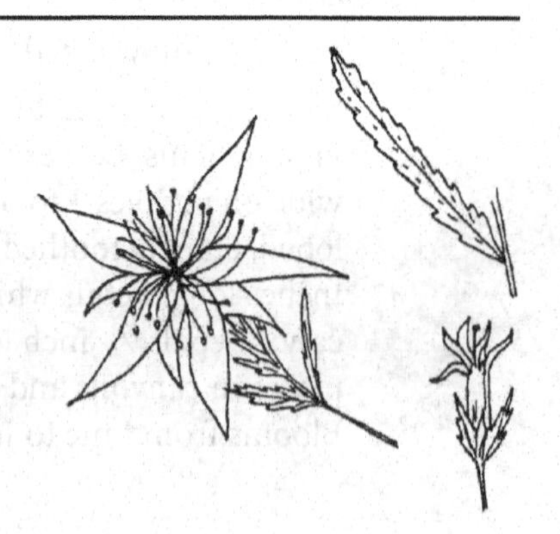

Loasaceae
Mentzelia laevicaulis (Hook.) Torr. & A. Gray

Giant Blazing Star

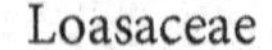

12. Leaves toothed or entire; not lobed

13. Upper leaf completely encircling stem. Flowers arising from center of upper leaf

Glabrous annual 4 to 12 inches high. Leaves mostly basal; these, long petioled and obovate. Cauline leaves 2; opposite, forming a circular disk which completely encircles the stem (perfoliate); ½ to 3 inches across. Flowers in a terminal raceme arising out of the perfoliate leaf. Petals 5; white, ¼ inch or less long. Sepals 5; stamens 5. Relatively common in the spring in moist shaded spots, particularly under the shade of oaks. Stem and leaves are delicious as a substitute for lettuce. Young plants are best. Look for miner's lettuce in early spring.

Montiaceae
Claytonia perfoliata Willd.

Miner's Lettuce

---

13. Upper leaf not completely encircling stem

14. Leaves in a basal rosette. Cauline leaves none or much reduced and more or less scale-like (Cauline leaves present, page 81)

15. Plants not succulent

16. Corolla lavender to red-violet. Lobes reflexed

Glandular pubescent perennial with basal leaves. Flowering scape 3 to 16 inches high. Leaves petiolate with spathulate, crisped, dentate blades; 2 to 4 inches long. Flowers nodding in a terminal few-flowered umbel. Corolla deeply 5-parted with reflexed lobes. Stamens 5, the filaments united. Corolla with white to red-violet lobes; yellow below, and maroon at base. Filaments black and anthers yellow. The entire flower appears quite delicate and resembles a flying bird. Quite common in early spring on moist, grassy slopes. Blooms from January to April.

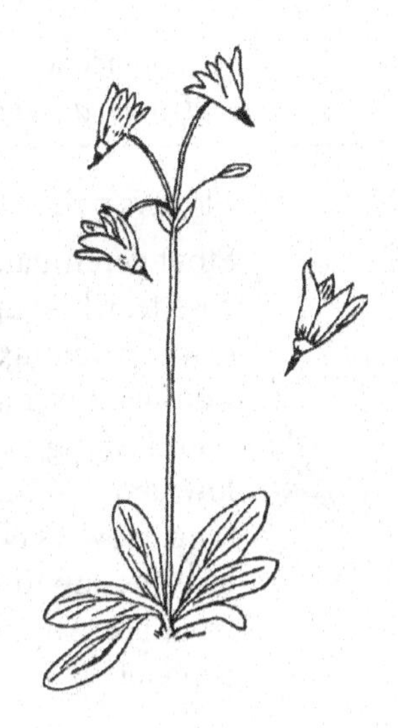

Primulaceae
Dodecatheon clevelandii Greene

Shooting Star

---

16. Petals white; not reflexed

Erect annual herb, 4 to 12 inches high with slightly glandular pubescent stems. Leaves are basal; petioled; ¼ to 2 inches long, ovate-oblong; with toothed margin. Flowers are delicate, with 5 white petals ⅛ to ¼ inch long. Sepals 5; purplish, 1/16 inch long, and reflexed after flowering. Stamens 10. Flowers occur in a few-flowered, open cluster. Fruit is a capsule. Occurs from the Santa Ana Mountains northward in shaded grassy areas. Blooms February to June.

Saxifragaceae
Micranthes californica (Greene) Small

California Saxifrage

---

15. Plants succulent

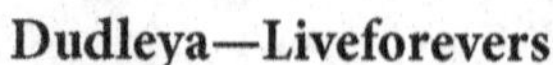

Dudleya—Liveforevers

16. Herbage densely covered with a white chalky powder

17. Flowers red. Plant of rocky cliffs and canyons

Robust plant densely covered with a white chalky powder. Basal leaves thick and large, in a rosette several inches across; spathulate-ovate; 3 to 10 inches long. Upper leaves much reduced, scale-like and clasping. Flowers in coiled racemes on terminal branches of an elongate flowering stem 16 to 32 inches high. Petals red; ½ to ¾ inch long; slightly fused at base and erect, not spreading at maturity. Fairly common on sides of rocky cliffs and canyons. Blooms from May to July.

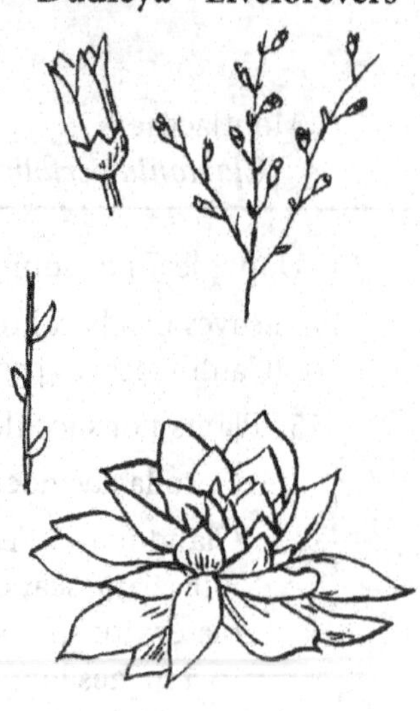

Crassulaceae
Dudleya pulverulenta (Nutt.) Britton & Rose

Chalk Dudleya

---

17. Flowers pale yellow. Plant of sea bluffs

Stout perennial, 4 to 14 inches high. Leaves in a basal rosette; white mealy to greenish, 1 to 2½ inches long; ovate to oblong; rounded beneath and flat above. Cauline leaves are short and scale-like, triangular and clasping, ⅜ to 1 inch long. Flowers are light yellow, about ½ inch long on stout pedicels up to ¼ inch long. Flowers occur in a 3 to 11 flowered cyme on branches up to 1½ inches long. Found along sea bluffs from Los Angeles County north. Blooms from May to September.

Crassulaceae
Dudleya farinosa (Lindl.) Britton & Rose

Bluff Lettuce

---

16. Herbage not densely covered with a white chalky powder

17. Flowers white

Succulent perennial 2 to 6 inches high with several basal leaves, 1 to 3 inches long, soon withering. Cauline leaves ¼ to 1 inch long; lanceolate to ovate. Stem and leaves glaucous and succulent. Flowers are white with red or purple tint, ¼ to ⅜ inch long, in an elongate inflorescence. Petals fused at the base, the lobes spreading. Flowers sweet odored. Blochman's dudleya is found near the coast from San Luis Obispo County south. It blooms from May to June. **RARE.**

Crassulaceae
Dudleya blochmaniae (Eastw.) Moran subsp. *blochmaniae*

Blochman's Dudleya

---

17. Flowers yellow to orange-red

18. Plant of sea bluffs; Los Angeles County and north

19. Petals pale yellow. Leaves ovate, more or less mealy

See *Dudleya farinosa*, page 78.

19. Petals bright yellow to red. Leaves lanceolate, shining

Stout succulent perennial, 4 to 14 inches high with leaves in a basal rosette. Leaves are thick, green; 2 to 8 inches long; oblong lanceolate. Flowers are yellow to orange-red, ⅜ to ⅝ inch long on stout pedicels ¼ inch long. Flowers occur on branches up to 4 inches long with up to 14 flowers. Grows on bluffs along coast from Los Angeles County north to Monterey. Blooms from April to July.

Crassulaceae
Dudleya caespitosa (Haw.) Britton & Rose

Sand/Sea Lettuce

---

18. Plant of dry rocky slopes

19. Flowers yellow (Flowers orange-red, page 81)

20. Corolla lobes spreading

Succulent perennial 4 to 14 inches high with leaves in a basal rosette. Basal leaves linear, terete, 1¼ to 6 inches long, soon withering. Cauline leaves ¼ to ¾ inch long, ovate-lanceolate and clasping at the base. Flowers yellow, 3/16 to ⅜ inch long; corolla lobes spreading. Calyx red. Pedicels ⅜ to ¾ inch long. Found on dry, stony soil from Los Angeles County south. Blooms from May to June. **RARE.**

Crassulaceae
Dudleya multicaulis (Rose) Moran

Many-Stemmed Dudleya

20. Corolla lobes erect

21. Rosette leaves ⅛ to ½ inch wide, withered in summer

22. Pedicels short, 1/16 to ⅛ inch long. Flowers pale yellow

Succulent perennial, 2 to 6 inches high with leaves in a basal rosette. Leaves linear to oblanceolate, ½ to 1½ inches long with acute apex. Cauline leaves lanceolate to ovate, ¼ to ½ inch long. Flowers pale yellow, ⅜ to ½ inch long on stout pedicels, 1/16 inch long. Flowers in a 5 to 10 flowered cyme 1 to 3 inches long. Found on bare, rocky slopes in southern Ventura County. Blooms from May to June. Leafless in the summer. **RARE.**

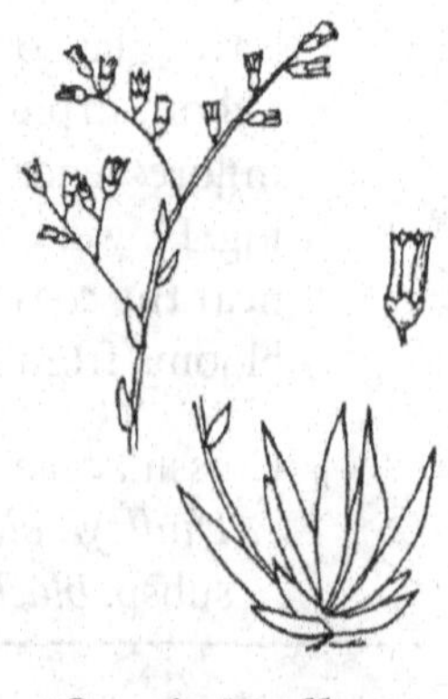

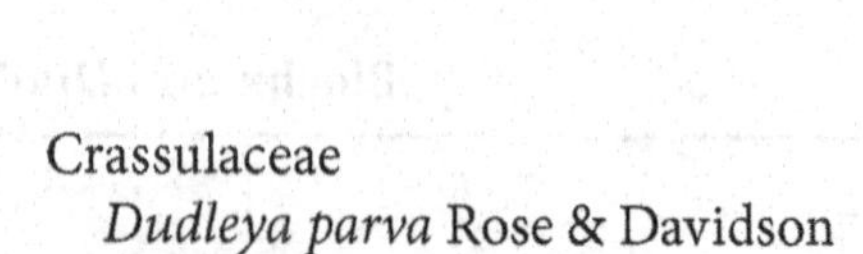

Crassulaceae
Dudleya parva Rose & Davidson

Conejo Dudleya

---

22. Pedicels longer, 3/16 to ½ inch long. Flowers bright yellow

Green, fleshy plant with basal leaves in a rosette. Leaves ⅝ to 1½ inches long and 3/16 to ½ inch wide; withering in the summer. Flowering stems 1½ to 4 inches high with 1 to 2 branches. Flowers on pedicels 3/16 to ½ inch long. Petals bright yellow, ⅜ to ½ inch long. Santa Monica Mountains dudleya grows in shaded canyons in the Santa Monica Mountains. Blooms from May to June. **RARE.**

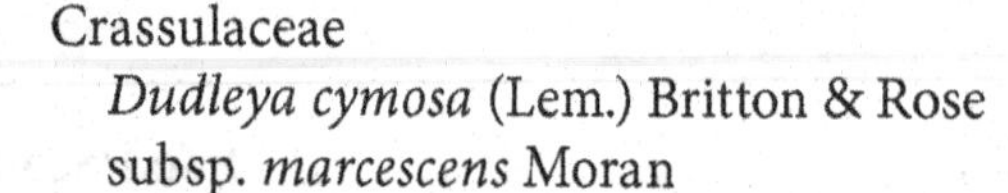

Crassulaceae
Dudleya cymosa (Lem.) Britton & Rose
subsp. *marcescens* Moran

Marcescent Dudleya

---

21. Rosette leaves ⅝ to 1 inch wide, not withered in summer

Green fleshy plant with a basal rosette of glabrous, ovate leaves ¾ to 2 inches long. Flower stalk 1½ to 6 inches high. Flowers in a coiled inflorescence on pedicels ¼ to ⅝ inch long. Petals bright yellow, ⅜ inch long, fused at base. Occurs in scattered localities in the Santa Monica Mountains. RARE.

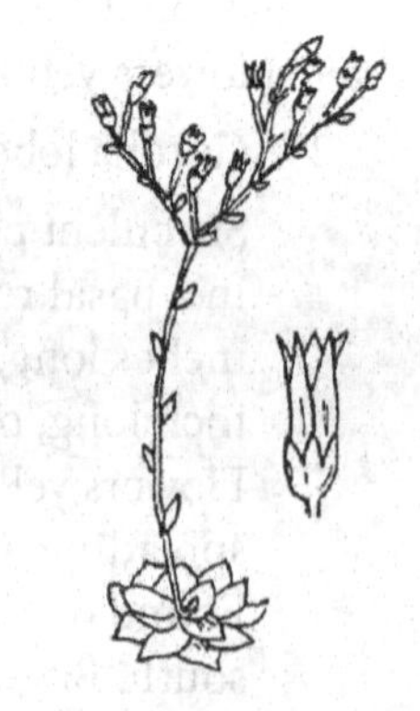

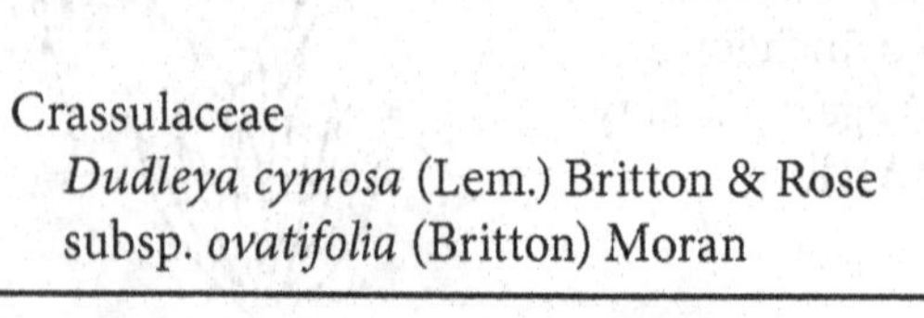

Crassulaceae
Dudleya cymosa (Lem.) Britton & Rose
subsp. *ovatifolia* (Britton) Moran

Santa Monica Dudleya

19. Flowers orange-red. Common

Green, fleshy perennial with flowering stems 8 to 24 inches high. Leaves in a basal rosette glaucous lanceolate; 2 to 8 inches long. Upper leaves much reduced, scale-like and clasping. Flowering stalk 8 to 24 inches high. Flowers in terminal, slightly coiled clusters on stout pedicels up to ⅜ inch long but mostly less than ¼ inch long. Petals waxy, orange-red; slightly fused at base; ⅜ to ⅝ inch long; erect at maturity. Very common plant on dry slopes. Abundant in Ventura County. Blooms from May to July.

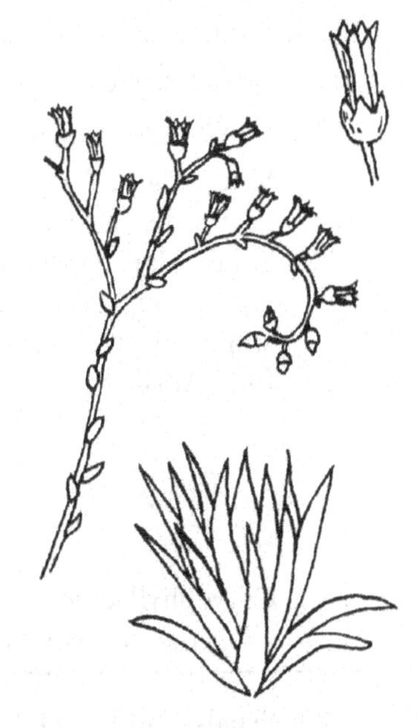

Crassulaceae
Dudleya lanceolata (Nutt.) Britton & Rose

Lance-Leaved Dudleya

---

14. Leaves not just basal; cauline leaves present

15. Leaves opposite or whorled (Leaves alternate, page 86)

16. Plant woody at base; a low bush. Flowers red-violet

Low bushy perennial with slender weak stems, repeatedly branched; sticky pubescent. Leaves ovate, more or less heart-shaped; opposite, entire. Flowers in a green calyx-like involucre; 1 flower per involucre. Petals none. Calyx colored, showy, and corolla-like; red-violet; 5-lobed; each lobe bi-lobed. Calyx about ½ inch long. Stamens 5, long exserted. Common flowering bush on dry slopes in the spring. Blooms from December to June.

Nyctaginaceae
Mirabilis laevis (Benth.) Curran
var. *crassifolia* (Choisy) Spellenb.

Wishbone Bush

---

16. Plant not woody at base

17. Plants found only on dunes or salt marshes along coast

18. Sepals fused; long cylindric. Flowers sessile

Low perennial herb, 6 to 12 inches high. Leaves opposite, entire, ovate below, narrower above; with axillary fascicles. Flowers solitary, sessile. Calyx cylindric, ¼ inch long. Petals lavender; ⅜ inch long. Found in salt marshes along coast. Blooms from June to October.

Frankeniaceae
Frankenia salina (Molina) I.M. Johnst.

Alkali Heath

---

18. Sepals not fused. Flowers short pedicelled

19. Leaves fascicled, appearing whorled

20. Sepals ⅛ to ¼ inch long. Petals white

Viscid glandular perennial 4 to 12 inches high. Leaves densely fascicled; linear; ⅜ to 1½ inches long. Stipules lanceolate, acuminate. Petals white, ⅛ to ¼ inch long. Sepals linear lanceolate, about the same length. Occurs in sandy soil near the beach. Blooms from April to July.

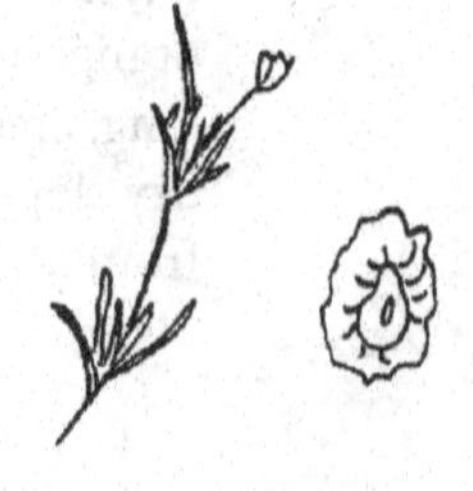

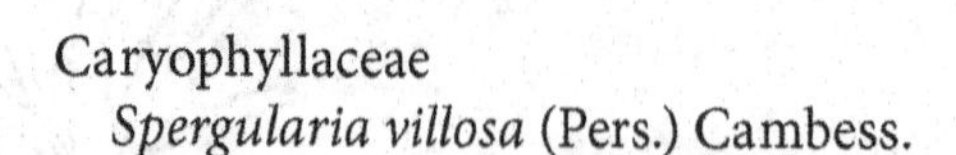

Caryophyllaceae
Spergularia villosa (Pers.) Cambess.

Hairy Sand-Spurrey

---

20. Sepals ¼ to ⅜ inch long. Petals pink

Prostrate to ascending perennial with glabrous or glandular pubescent stems 3 to 12 inches long. Leaves, linear, densely fascicled; ⅜ to 1½ inches long. Stipules triangular acuminate, ¼ to ⅜ inch long. Petals pink, about ¼ inch long; sepals slightly longer. Stamens 10. Found in salt marshes and in salty areas along the beach. Blooms most of the year.

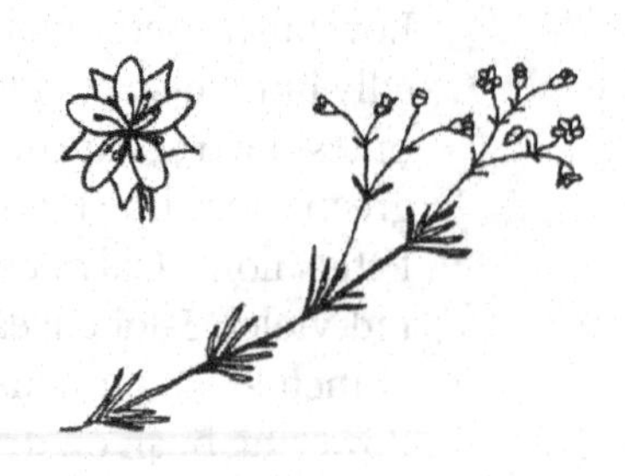

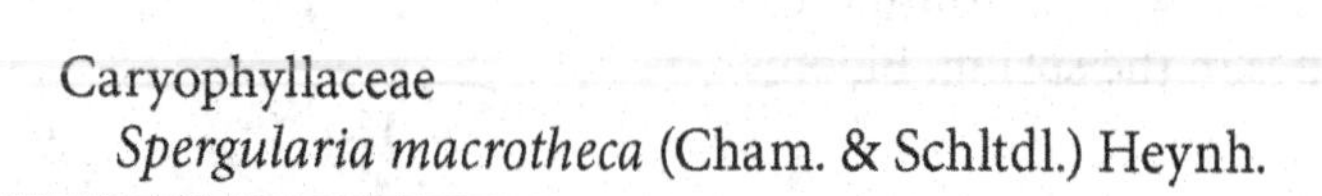

Caryophyllaceae
Spergularia macrotheca (Cham. & Schltdl.) Heynh.

Sticky Sand-Spurrey

---

19. Leaves not fascicled

20. Sepals ⅛ to ¼ inch long

Annual, 2 to 12 inches high, glabrous below and glandular pubescent above. Leaves opposite with deltoid, membranous stipules. Leaves linear, ⅜ to ¾ inch long. Flowers white to pink, shorter than the sepals. Sepals ⅛ to ¼ inch long. Occurs along the beach and in alkaline places. Blooms from April to September.

Caryophyllaceae
Spergularia bocconi (Scheele) Graebn.

Boccone's Sand-Spurrey

---

20. Sepals ¼ to ⅜ inch long

Low, spreading annual; glandular pubescent; 2 to 12 inches high. Leaves fleshy, linear, mostly not fascicled; ¾ to 1½ inches long; opposite with scarious stipules. Flowers in terminal racemes. Sepals 5; ovate, ¼ to ⅜ inch long. Petals 5; white to pink, slightly shorter than the sepals. Stamens 2 to 5. Common plant of salty and alkaline areas along the coast. There are other species of *Spergularia* that may also be found that are similar to this one. Blooms April to September.

Caryophyllaceae
Spergularia marina (L.) Besser

Saltmarsh Sand-Spurrey

---

17. Plants not found only on dunes or salt marshes along the coast

18. Leaves whorled

Erect annual with glabrous to glandular pubescent stems 4 to 16 inches high and whorled leaves. Leaves are linear; ⅜ to 1¼ inches long; and less than 1/16 inch across. Flowers are white with 5 petals and 5 sepals each about 3/16 inch long. Stamens are 5 to 10; styles 5. Flowers occur in an open cluster. Fruit is an ovoid capsule. Stickwort has become naturalized in fields and vacant lots. Blooms most of the year.

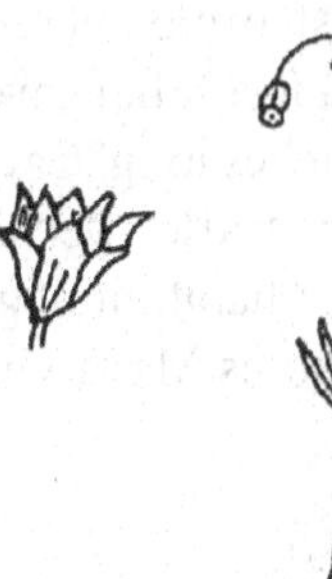

Caryophyllaceae
Spergula arvensis L.

Stickwort

---

18. Leaves opposite; not whorled

19. Petals small (less than ¼ inch long); 2 cleft; shorter than sepals

20. Flowers solitary in leaf axils. Plant glabrous except for line of hairs between nodes

Slender, weak-stemmed annual with trailing stems. Leaves opposite, entire, ovate; mostly ¼ to 1 inch long. Flowers small, solitary in leaf axils. Sepals 5; 3/16 inch long. Petals 4 to 5, white; shorter than the sepals, bifid at apex. Common weed in moist shaded areas. Blooms from February to September.

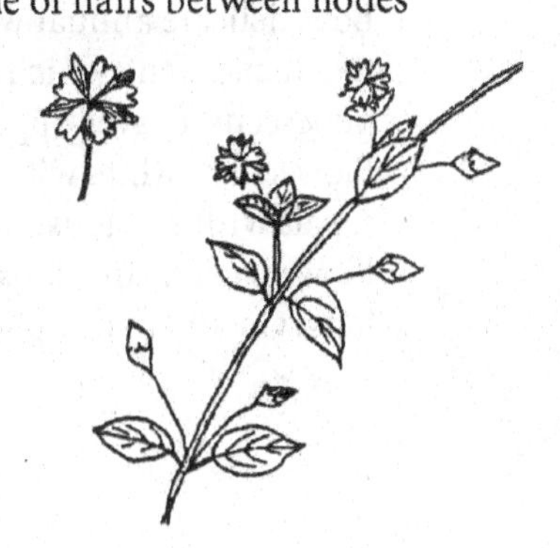

Caryophyllaceae
Stellaria media (L.) Vill.

Common Chickweed

---

20. Flowers in terminal clusters. Plant sticky, pubescent

Viscid pubescent annual 4 to 12 inches high. Leaves are opposite, narrow, ovate, ⅜ to 1 inch long with obtuse margins. Flowers have 5 white, 2-cleft petals, the same length or shorter than the sepals. Sepals 5; $^{3}/_{16}$ inch long with scarious margins. Fruit a slightly curved capsule ¼ to ⅜ inch long. Common plant of waste places. Blooms February to May.

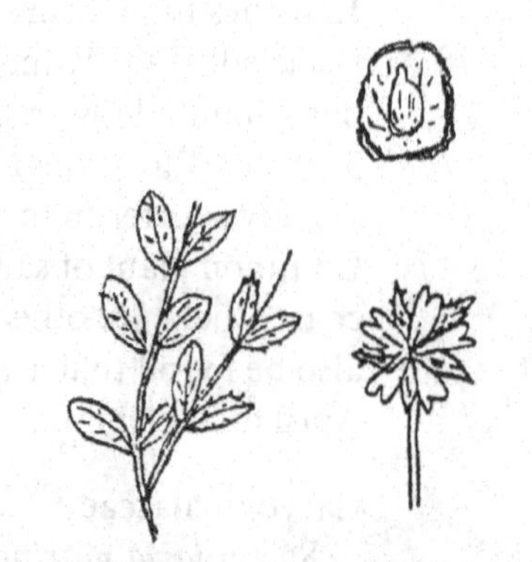

Caryophyllaceae
Cerastium glomeratum Thuill.

Sticky Mouse-Ear Chickweed

---

19. Petals not 2-cleft; longer than the sepals

20. Flowers red, showy. Petals cleft into 4 linear lobes

Erect, weak-stemmed perennial, glandular pubescent; mostly 1 to 4 feet tall. Leaves opposite, linear lanceolate, 2 to 4 inches long. Calyx urn-shaped; up to ¾ inch long. Petals 5; scarlet; deeply 4-cleft. Corolla ½ to ¾ inch across. Inhabitant of grassy or dry, brushy slopes and hills. Blooms May to July.

Caryophyllaceae
Silene laciniata Cav.
subsp. *laciniata*

Mexican Pink

---

20. Flowers not red. Petals not deeply cleft

21. Leaves filiform

Low, delicate annual with slightly matted stems 2 to 8 inches long which are more or less glandular pubescent. Leaves opposite, filiform; ¼ to ⅝ inch long. Flowers are white with 5 petals, ¼ inch long, and 5 sepals with scarious margins. Stamens 10; styles 3. Flowers occur in a loose cyme on pedicels ¼ to 1 inch long. Occurs in dry places. Blooms from April to June.

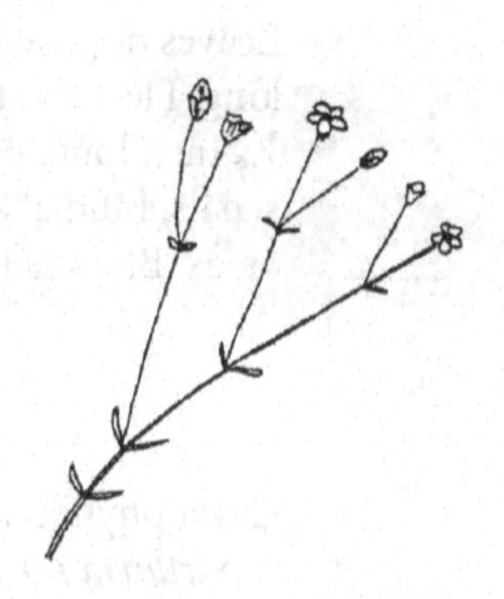

Caryophyllaceae
Minuartia douglasii (Torr. & A. Gray) Mattf.

Douglas' Stitchwort

---

21. Leaves broader

22. Flowers yellow. Plant prostrate, fleshy

Prostrate annual with fleshy, more or less reddish stems 4 to 8 inches long. Leaves somewhat thick and fleshy, obovate to spathulate, up to 1 inch long. Petals 5; pale yellow; ⅛ to ¼ inch across. Sepals 2; stamens 7 to 20. Common weed in waste places. Flowers from May to September. Leaves and young stem are edible. They may be eaten raw or boiled for a potherb.

Portulacaceae
Portulaca oleracea L.

Purslane

---

22. Flowers not yellow. Plant not fleshy

23. Plant low, spreading. Flowers salmon-colored, open only under fair skies

Low, spreading annual with flat-lying branched stems. Leaves opposite, sessile, entire, ovate; ¼ to ¾ inch long. Flowers pedicelled in leaf axils. Pedicels about ½ inch long. Corolla 5-parted, salmon-colored, ⅜ inch across. Stamens 5. Common weed, often in moist places. Blooms from March to July. Flowers open only under blue sky. This plant is included in the literature as a **TOXIC PLANT** and is believed to have caused the death of horses.

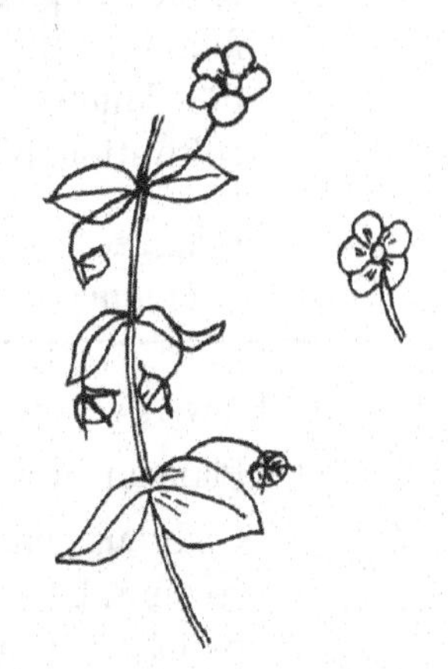

Myrsinaceae
Anagallis arvensis L.

Scarlet Pimpernel

---

23. Plant erect. Flowers pale pink to white. Sepals fused, urn-shaped

Erect annual, glandular hirsute, with stems 4 to 16 inches high. Leaves opposite, oblanceolate with a mucronate tip. Flowers in a one-sided inflorescence. Calyx urn-shaped; villous; 5-toothed, 10-nerved; becoming inflated in age; ¼ to ⅜ inch long; constricted at top. Petals 5, clawed, pink to white, twisted slightly, resembling a windmill. Stamens 10. Common weed in lawn and garden. Blooms from February to June.

Caryophyllaceae
Silene gallica L.

Windmill Pink

---

15. Leaves alternate

16. Flowers red to red-violet

17. Petals 3/16 to 3/8 inch long. Leaves somewhat fleshy

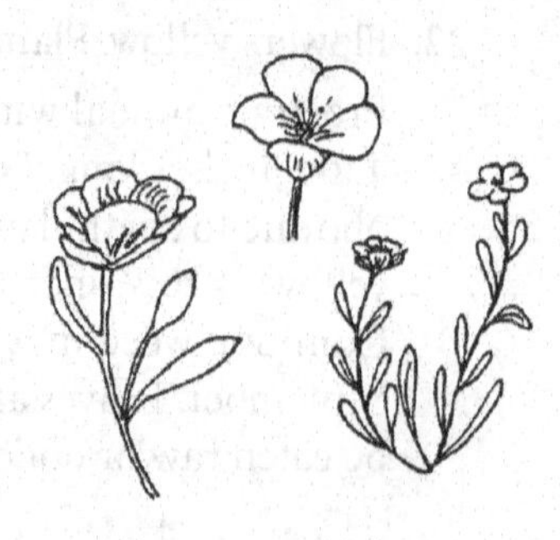

Low, glabrous annual, 4 to 16 inches high. Leaves alternate, entire, linear to oblanceolate, 3/4 to 3 inches long, somewhat fleshy. Flowers magenta or red-violet, 3/8 to 3/4 inch across. Sepals 2; petals 5. Fairly common in early spring on grassy slopes and flats. Blooms from February to May.

Montiaceae
Calandrinia ciliata (Ruiz & Pav.) DC.

Red Maids

---

17. Petals 1/2 to 3/4 inch long. Leaves not fleshy

Glabrous annual 8 to 20 inches high. Leaves alternate, lanceolate, 3/8 to 3/4 inch long, sessile, entire. Flowers showy, bright red, on pedicels 3/4 to 2 3/4 inches long. Petals 5, 5/8 to 3/4 inch long. Red flax is not native to our area but may be found frequently as it does grow well and is often present in wildflower packets. It is also an occasional escape from cultivation. Blooms from April to June.

Linaceae
Linum grandiflorum Desf.

Red Flax

---

16. Flowers yellow

17. Plant aquatic; in pools or slow moving streams

Glabrous perennial growing in pools and slow moving streams. Leaves oblong, entire, 3/8 to 1 1/2 inches long on petioles 1 inch or more long. Flowers yellow, on pedicels 3/8 to 1 1/2 inches long. Petals 5; stamens 10. Flowers 3/8 to 5/8 inch long. Blooms from May to October.

Onagraceae
Ludwigia peploides (Kunth) P.H. Raven

Yellow Water Primrose

---

17. Plant not aquatic

18. Leaves linear

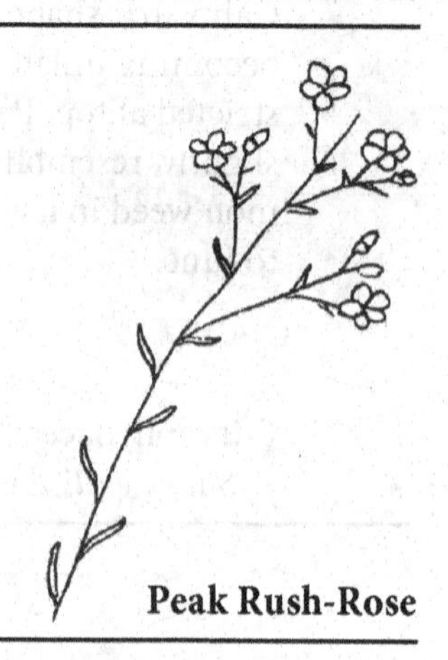

Bushy plant with stems 8 to 12 inches high. Leaves linear, dropping early; mostly 1/2 to 1 inch long. Flowers in a terminal panicle. Petals 5; 1/4 inch long, yellow. Sepals 5, the two outer ones linear; the three inner ones ovate, 3/16 inch long. Common in sandy areas near the coast. Blooms from March to June.

Cistaceae
Helianthemum scoparium Nutt.

Peak Rush-Rose

---

18. Leaves lanceolate to spatulate
19. Flowers large. Petals 1 to 2½ inches long
See *Mentzelia laevicaulis*, page 76.
19. Flowers small. Petals less than ¼ inch long
20. Plant prostrate. Leaves entire
See *Portulaca oleracea*, pages 56 and 85.
20. Plant erect. Leaves toothed
See *Mentzelia micrantha*, page 76.

9. Petals less than 5
10. Petals 4 (Petals 3, page 101)
11. Leaves all basal. Flowers minute in a dense, terminal spike or raceme
12. Flowers in a terminal spike
See *Plantago*, page 47 and 48.
12. Flowers in a terminal raceme
See *Athysanus*, page 47.
11. Leaves and flowers not as above
12. Leaves whorled
See *Galium*, page 59.
12. Leaves not whorled; opposite or alternate
13. Stamens numerous; more than 10
14. Flowers yellow to orange. Leaves dissected into linear segments

Glaucous herb; stems up to 2 feet long. Leaves several times dissected into linear segments. Flowers showy, bright yellow or orange. Sepals 2; united, falling as petals enlarge. Petals 4; ¾ to 2¼ inches long. Stamens numerous. Pistil 1. Common plant on grassy slopes, sometimes covering large areas and producing splashes of color. The western Mojave Desert around Lancaster is particularly noted for its spring poppy display. Blooms from February to September. Summer flowers are usually smaller and yellow rather than orange. The **Tufted Poppy,** *Eschscholzia caespitosa* Benth., resembles *E. californica*, except that the petals are only ⅜ to 1 inch long. Blooms from March to June. *E. caespitosa* is generally less common than *E. californica*.

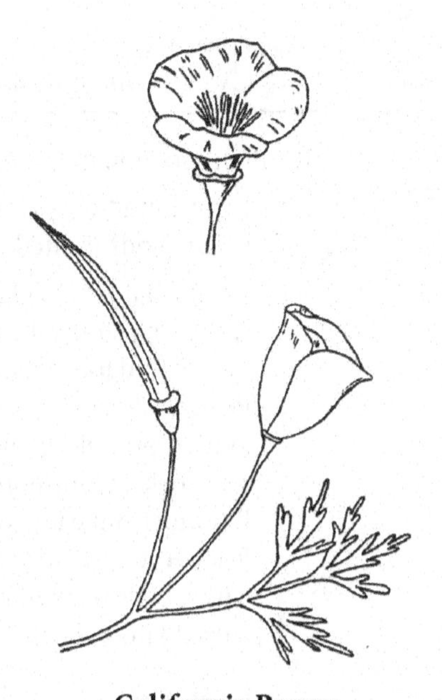

Papaveraceae
Eschscholzia californica Cham.

California Poppy

14. Flowers orange-red. Leaves pinnately divided or lobed

Erect slender annual 1 to 2 feet high with yellow or milky juice. Leaves pinnately divided into rounded, lobed segments. Flowers terminal; showy; brick-red. Sepals 2, hairy, ⅜ inch long. Petals 4; ½ to ¾ inch long. Stamens numerous. Occurs in burned areas and disturbed places in chaparral woods. Not common. Blooms April to May.

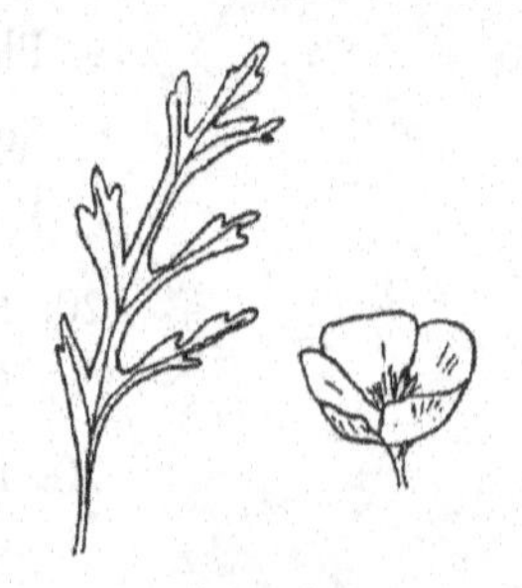

Papaveraceae
Papaver californicum A. Gray

Fire Poppy

---

13. Stamens less than 10

14. Stamens 6; 4 long and 2 short (Stamens 8, page 95) **Brassicaceae—Mustard Family**

15. Fruit (pod) elongated; two or more times longer than wide (Pod orbicular, page 93)

16. Leaves compound or pinnatifid dissected (Leaves simple, page 93)

17. Plant aquatic; in water

Aquatic perennial with weak, watery stems; prostrate or partially erect. Leaves pinnately dissected; ¼ to 4 inches long. Flowers small, in a terminal raceme. Petals white, mostly ⅛ inch long. Fruits curved, about ⅜ inch long at maturity. Occurs in quiet water or on wet stream banks. Blooms from March to November. Water cress is well known for its edibility as a salad plant. It can also be boiled as a potherb. Do **NOT** eat plants found in polluted water because **TOXIC** elements can become concentrated in the plant tissue.

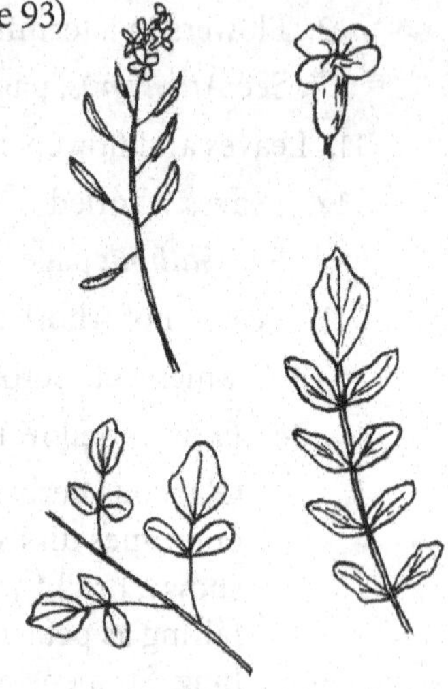

Brassicaceae
Nasturtium officinale W.T. Aiton

Water Cress

---

17. Plant not aquatic; not floating in water

18. Plant found only on sand dunes. Herbage more or less fleshy. Pods jointed

Somewhat fleshy, glabrous annual with flat-lying, spreading stems. Leaves pinnately parted or divided into round lobes. Flowers in a terminal raceme. Petals pale violet; ¼ inch long. Fruit a swollen two-jointed pod; fleshy when young but becoming hard and corky. Common on sand dunes along the coast. Blooms from May to September. The **European Sea Rocket**, *Cakile maritima* Scop., is more common, with petals pale violet; ¼ to ½ inch long and fruit with 2 lateral horns.

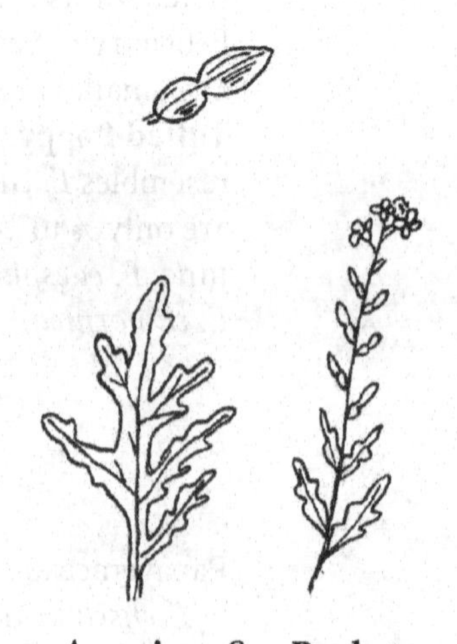

Brassicaceae
Cakile edentula (Bigelow) Hook.

American Sea Rocket

---

18. Plant not found only on sand dunes. Herbage not fleshy. Pods not jointed

19. Flowers more or less showy. Petals ⅜ to ¾ inch long

20. Plant of shaded woods. Leaves 3 to 5 foliate. Petals ⅜ to ¾ inch long

Erect, glabrous perennial 4 to 16 inches high with leafless stems below. Leaves mostly 3 to 5 foliate; leaflets ovate to lanceolate, toothed, lobed, or entire. Flowers showy, white to pale pink, in a many-flowered raceme. Petals ⅜ to ¾ inch long. Fruit a linear pod ¾ to 2 inches long. Common plant of shaded woods, blooming from February to May.

Brassicaceae
Cardamine californica (Nutt.) Greene

Milk Maids

---

20. Plant of dry open areas. Leaves pinnatifid or pinnately parted. Petals ½ to ¾ inch long

21. Plant of desert areas. Fruit a thin, linear pod, 1 to 3 inches long. Flowers yellow, not veined, in a dense, terminal raceme, 4 to 24 inches long

Tall bushy plant 1½ to 5 feet high. Lower leaves pinnatifid; upper leaves divided or entire. Flowers in a showy terminal raceme. Petals lemon-yellow; ½ to ⅝ inch long. Sepals 4. Pod linear, spreading; 1 to 3 inches long, on a thin stipe ⅜ to 1 inch long. More common in dry desert areas or in selenium-bearing soil. Blooms from April to September.

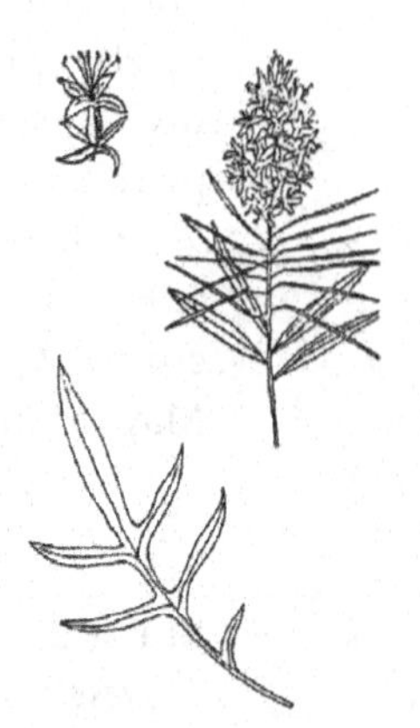

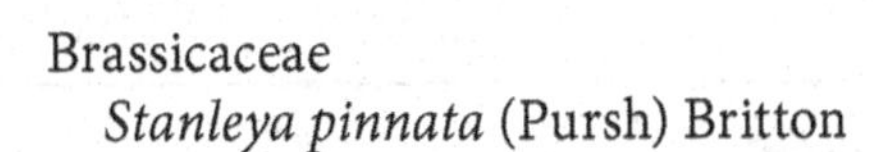

Brassicaceae
Stanleya pinnata (Pursh) Britton

Prince's Plume

---

21. Common weed. Fruit carrot-shaped, ¼ inch thick, ¾ to 1½ inches long. Flowers white, yellow, pink; variously veined in a shorter, open inflorescence

Erect, branched plant 1 to 4 feet high. Lower leaves pinnately parted with a large rounded terminal segment. Upper leaves simple, ovate, toothed. Flowers showy in a terminal raceme. Petals ½ to ¾ inch long; clawed; yellow, pink, lavender, or white with purple veins. Fruits ¾ to 1½ inches long, ascending; ¼ inch thick at maturity. Common weed in waste places. Everywhere abundant. Blooms from February to July.

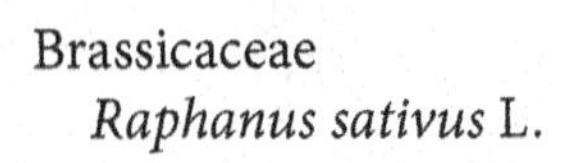

Brassicaceae
Raphanus sativus L.

Radish

---

19. Petals less than ⅜ inch long

20. Leaves 2 to 3 times pinnatifid. Pods more or less curved

21. Pods short, less than ½ inch long; oblong

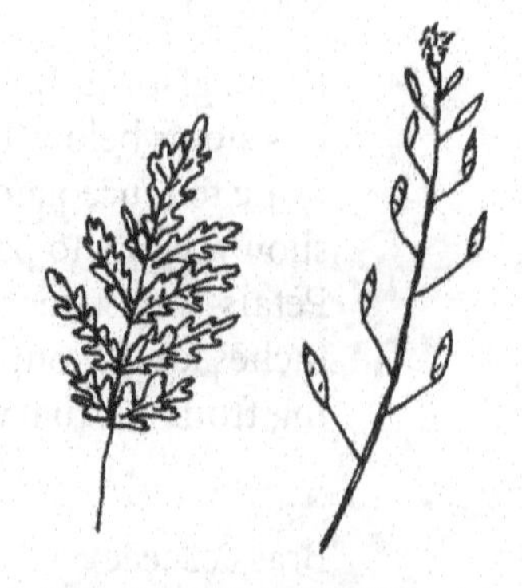

Pubescent annual 4 to 24 inches high. Leaves very finely dissected into narrow divisions. Flowers minute, in a terminal raceme. Petals yellow, less than ⅛ inch long. Fruits arcuate, less than ½ inch long; about 1/16 inch wide. Common in dry sandy places, including the Mojave and Colorado Deserts. Blooms from March to June.

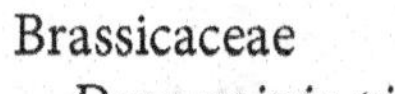

Brassicaceae
Descurainia pinnata (Walter) Britton

Western Tansy Mustard

---

21. Pods ⅜ to 1¼ inches long; linear

Erect annual, 10 to 24 inches high with stellate pubescent stems. Leaves 1 to 3½ inches long in a basal rosette; 2 or 3 times pinnate compound into linear divisions. Flowers in a raceme on somewhat spreading pedicels ¼ to ½ inch long. Petals 4; greenish yellow; 1/16 inch long. Sepals 4; same length as petals. Stamens 6. Fruit a linear silique; ⅜ to 1¼ inches long; often curved. Occasional weed of dry places. Blooms from May to August.

Brassicaceae
Descurainia sophia (L.) Prantl

Flix Weed

---

20. Leaves once pinnatifid; not 2 or 3 times pinnatifid

21. Pods reflexed

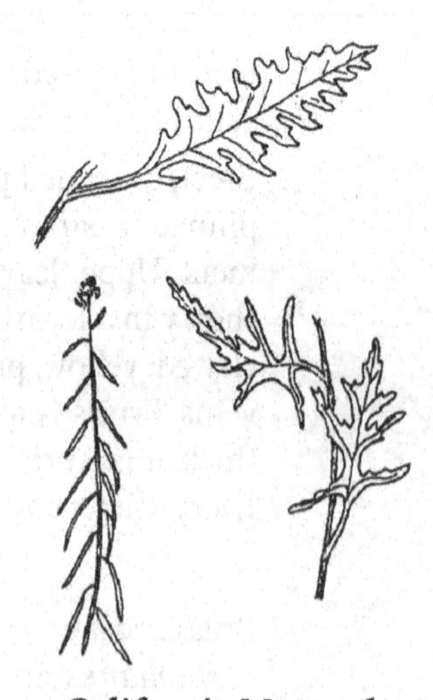

Erect annual, hirsute at the base and 8 to 40 inches high. Leaves are irregularly pinnatifid or sinuate dentate; 1 to 4½ inches long; oblong to oblanceolate. Upper leaves are reduced. Flowers occur in a raceme on pedicels ⅛ inch long which are turned down or reflexed after flowering. Petals 4; yellowish white; ¼ inch long. Sepals 4; yellowish green. Fruit a linear silique 2½ inches long; reflexed. Occurs on slopes and in burned-over areas. Blooms from March to June.

Brassicaceae
Caulanthus lasiophyllus (Hook. & Arn.) Payson

California Mustard

---

21. Pods spreading or ascending

22. Fruits with a beak

23. Stems hirsute; perennial

Erect biennial or perennial with more or less hirsute stems 1 to 3 feet high and bush-like. Basal leaves are $1\frac{1}{4}$ to 4 inches long and pinnatifid with a large terminal lobe. Upper leaves are smaller; toothed or lobed. Flowers are pale mustard yellow; $\frac{1}{2}$ inch long and occur in a raceme on short pedicels appressed to the stem. Fruit a silique up to $\frac{1}{2}$ inch long with a short beak. Siliques erect, closely appressed to stem. Very common weed in fields and waste places. Blooms from May to October. Young leaves and flowers are edible. Leaves may be cooked or eaten in a salad.

Brassicaceae
Hirschfeldia incana (L.) Lagr.-Fossat

Shortpod Mustard

---

23. Stems glabrous or with a few scattered hairs. Annual

24. Beak $\frac{3}{8}$ to $\frac{5}{8}$ inch long

Erect, glabrous annual, 1 to 4 feet high. Lower leaves are pinnatifid or lobed; 4 to 8 inches long; petioled. Upper leaves are sessile; auriculate clasping; glabrous and entire. Flowers have 4 yellow petals, $\frac{1}{4}$ inch long and occur on pedicels $\frac{3}{4}$ inch long. Sepals 4; yellowish. Fruit a linear silique $\frac{3}{4}$ to 2 inches long with a stout beak up to $\frac{1}{2}$ inch long. Common weed of field and waste places. Blooms from January to May.

Brassicaceae
Brassica rapa L.

Field Mustard

---

24. Beak $\frac{3}{16}$ inch long

Annual, erect and sparingly pubescent; $1\frac{1}{2}$ to 8 feet tall, not bush-like. Lower leaves large, coarse, deeply pinnatifid with a large terminal lobe. Upper leaves reduced. Flowers somewhat showy in a terminal raceme; only the top portion flowering at one time, however. Petals lemon yellow, clawed, averaging $\frac{1}{4}$ inch long. Fruit an elongate, linear pod with a short beak. Pods appressed against the stem; $\frac{1}{4}$ to $\frac{3}{4}$ inch long. A very common weed, often coloring entire fields yellow in the late spring. Blooms from April to July.

Brassicaceae
Brassica nigra (L.) W.D.J. Koch

Black Mustard

---

22. Fruits without a beak

23. Fruits 1 to 4 inches long, spreading

24. Petals ¼ to ⅜ inch long

25. Upper leaves glabrous; pinnatifid

Erect, glabrous annual 1½ to 3½ feet high. Lower leaves pinnatifid, 4 to 6 inches long with a large, triangular terminal lobe. Upper leaves reduced and pinnatifid into linear divisions. Petals 4; yellowish white; ¼ inch long. Fruits 2 to 4 inches long; linear; spreading. Common weed. Blooms from May to July.

Brassicaceae
Sisymbrium altissimum L.

Tumble Mustard

25. Upper leaves pubescent; entire to hastate

Erect annual or biennial with pubescent or more or less hirsute stems, 8 to 24 inches high. Lower leaves pinnatifid; 3 to 5 inches long. Upper leaves entire to hastate lobed. Petals 4; yellow; ¼ inch long on pedicels up to ⅜ inch long. Fruit 1½ to 3½ inches long, ascending to spreading. Weed from Ventura County south to San Diego. Blooms in May.

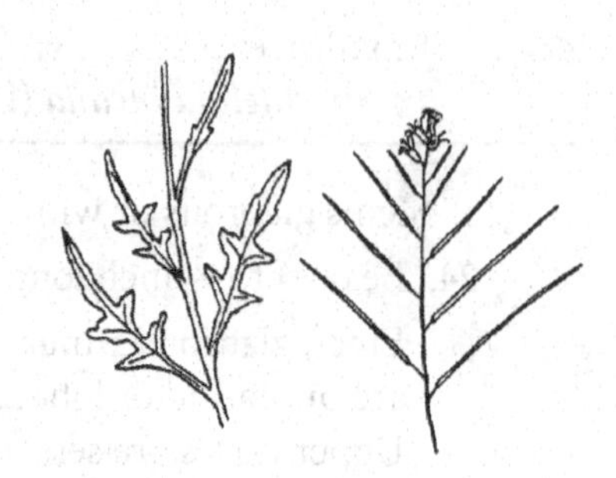

Brassicaceae
Sisymbrium orientale L.

Oriental Mustard

24. Petals ⅛ inch long

Glabrous annual ½ to 3 feet tall. Leaves deeply pinnatifid with a larger terminal lobe. Flowers pale yellow in a terminal raceme, only top flowers blooming at a time. Petals ⅛ inch long. Fruit an elongate, linear pod, 1½ to 1¾ inches long; ascending or pointing outward at maturity. Fairly common weed of spring in orchards or fields. Not nearly as common as black mustard, however. Blooms from January to April.

Brassicaceae
Sisymbrium irio L.

London Rocket

23. Fruits less than 1 inch long; appressed to stem

Erect annual with more or less hirsute stems 8 to 40 inches high. Basal leaves pinnatifid, 2 to 4 inches long. Upper leaves hastate clasping; hirsute. Petals 4; yellowish; ⅛ inch long on short pedicels. Flowers in a long, narrow raceme. Fruits 1 inch or less, closely appressed to stem. Blooms from April to July. Common weed of orchards and fields. Young plants may be cooked as a potherb.

Brassicaceae
Sisymbrium officinale (L.) Scop.

Hedge Mustard

16. Leaves simple; toothed to entire

17. Flowers orange-yellow

Erect, mostly unbranched plant ¾ to 3 feet high. Basal leaves long lanceolate, few toothed; 1½ to 6 inches long. Upper leaves reduced and mostly entire. Flowers in terminal racemes, orange-yellow; showy. Petals clawed, ½ to ¾ inch long. Fruit linear, 2 to 4 inches long; ascending. Common plant on dry slopes. Blooms from March to July.

Brassicaceae
Erysimum capitatum (Hook.) Greene

Western Wallflower

---

17. Flowers purple

Perennial with pubescent herbage; erect, mostly unbranched; 1 to 3 feet tall. Basal leaves lanceolate, dentate with obtuse apex. Upper leaves entire, oblanceolate, sessile with sagittate base. Flowers purple, in a short terminal raceme. Petals ¼ to ½ inch long. Fruits linear; 2 to 4½ inches long, ascending. Occasional on dry slopes. Blooms from February to May.

Brassicaceae
Boechera californica (Rollins) Windham & Al-Shehbaz

California Rockcress

---

15. Pods orbicular or heart-shaped. Very little longer than wide

16. Pods heart shaped

Erect, pubescent annual, 8 to 20 inches high. Basal leaves form a rosette; pinnatifid dissected. Upper leaves becoming reduced and ultimately entire; lanceolate, sessile with an auricled base. Flowers small, in a terminal raceme. Petals white, $\frac{1}{16}$ inch long. Fruits flattened, heart-shaped or purse-shaped; about ⅜ inch long and ¼ inch wide. Common weed on dry or disturbed soil. Blooms most of the year.

Brassicaceae
Capsella bursa-pastoris (L.) Medik.

Shepherd's Purse

---

16. Pods not heart-shaped

17. Fruits with a winged margin

See *Thysanocarpus*, page 58.

17. Fruits without a winged margin

18. Leaves linear, entire

Low perennial, 2 to 10 inches high. Leaves linear, entire, about ½ inch long. Flowers small, white, in a terminal raceme. Fruit orbicular, ⅛ inch long. Often cultivated and occurs as an escape. Also occurs along the beach. Blooms most of the year.

Brassicaceae
Lobularia maritima (L.) Desv.

Sweet Alyssum

---

18. Leaves toothed or pinnatifid

19. Pods notched at apex

Annual, 2 to 16 inches high. Lower leaves pinnately divided; the upper leaves pinnatifid or entire. Flowers small, white, in a terminal raceme. Petals $^{1}/_{16}$ inch long or less. Fruit orbicular, up to ¼ inch long at maturity; notched at apex. Common plant on open slopes. Blooms from January to April.

Brassicaceae
Lepidium nitidum Nutt.

Shining Peppergrass

---

19. Pods not notched at apex

Pubescent perennial 12 to 16 inches high; much branched. Leaves ovate, sharply toothed, 1 to 1½ inches long. Upper leaves clasping with an auriculate base. Flowers small, white, in a terminal raceme. Petals ⅛ inch long. Fruits ovoid, acute at apex, with a persistent style. Fairly common weed in some localities, but not nearly as abundant as the above. Blooms from March to June.

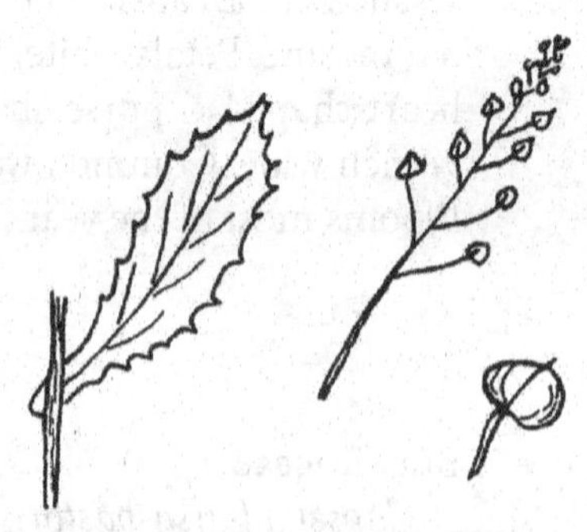

Brassicaceae
Lepidium draba L.

Heart-Podded Hoary Cress

---

14. Stamens 8; 4 long and 4 short. Ovary inferior **Onagraceae—Evening-Primrose Family**

15. Sepals and petals scarlet. Plant woody at base

16. Leaves gray, entire. Largest leaves not more than ½ inch wide

Slightly woody perennial with reddish stems; pubescent but not very glandular. Leaves gray, linear to linear lanceolate; entire; no more than 1/12 inch wide; densely fascicled. Flowers scarlet, averaging 1 inch long. Both sepals and petals scarlet; 4 in number. Stamens 8; exserted. Common plant of late summer on dry slopes of chaparral.

Onagraceae
Epilobium canum (Greene) P.H. Raven
subsp. *canum*

Hoary Fuchsia

---

16. Leaves green, toothed. Largest leaves more than ½ inch wide

Slightly woody, glandular pubescent perennial; 12 to 36 inches high. Leaves green or gray-green; sometimes reddish; lanceolate to ovate, dentate and often fascicled. Flowers similar to the above. Petals 2-cleft; scarlet. Flowers mostly ¾ to 1¼ inches long. Common plant in dry areas. Blooms from August to October.

Onagraceae
Epilobium canum (Greene) P.H. Raven
subsp. *latifolium* (Hook.) P.H. Raven

California Fuchsia

---

15. Sepals and petals not both scarlet

16. Flowers yellow or cream
(Flowers pink, blue, violet, or white, page 97)

17. Flowers large, showy. Petals 1 to 1½ inches long; light-yellow

Erect, rather stout herb, 1½ to 4 feet high with reddish stem; densely pubescent. Basal leaves oblanceolate, sinuate entire, short petiolate or sessile. Flowers large, showy, yellow, fading reddish orange. Petals 4; 1 to 1½ inches long. Sepals 4; reflexed; ¾ to 1½ inches long. Ovary inferior, longitudinally ribbed and densely pubescent. Occurs in moister areas of the inland chaparral. Blooms from June to September. **Hooker's Evening-Primrose,** *Oenothera elata* Kunth subsp. *hookeri* (Torr. & A. Gray) W. Dietr., occurs along the coast.

Onagraceae
Oenothera elata Kunth
subsp. *hirsutissima* (S. Watson) W. Dietr.

Hairy Evening-Primrose

---

17. Flowers smaller. Petals not more than 1 inch long; bright yellow

18. Plant found only on sand dunes along the coast. Herbage silvery-gray

Perennial with prostrate stems; densely pubescent with silvery-gray herbage. Upper leaves sessile, ovate, entire. Flowers, sessile, in leaf axils. Petals bright yellow, ½ to 1 inch long. Sepals reflexed; 3/16 to ⅜ inch long. Fruits densely pubescent; 4-angled; becoming coiled at maturity; ½ to ⅞ inch long. Common plant on sand dunes along the coast. Blooms from April to August.

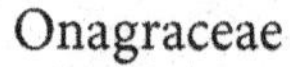

Onagraceae
Camissoniopsis cheiranthifolia (Spreng.) W.L. Wagner & Hoch

Beach Evening-Primrose

---

18. Plant not found only on sand dunes along the coast

19. Petals more than 3/16 inch long

20. Plant prostrate

Prostrate annual; stems radiating outward and often reddish. Leaves lanceolate, pubescent, toothed, or entire, ½ to 4½ inches long. Basal leaves petiolate; upper leaves entire. Flowers in axils of leaves. Petals 4; bright yellow with a dark maroon spot at base; ¼ to ⅝ inch long. Sepals 4; reflexed. Fruit an elongate linear pod, 4-angled and pubescent; usually curved or contorted at maturity. Common inhabitant of sandy areas. Blooms from March to June.

Onagraceae
Camissoniopsis bistorta (Torr. & A. Gray) W.L. Wagner & Hoch

California Sun Cup

---

20. Plant erect

Tall, erect, glabrous, somewhat coarse annual 1 to 3 feet high. Basal leaves in a rosette, pinnatifid, dying early. Upper leaves lanceolate, toothed, few and far apart. Flowers yellow, drying orange or reddish. Sepals 4; reflexed; about ¼ inch long. Petals ¼ to ⅝ inch long. Pods long linear, 1 to 4 inches long; not curved; bent downward at maturity. Fairly common plant in dry and disturbed places of the chaparral. Blooms from April to May.

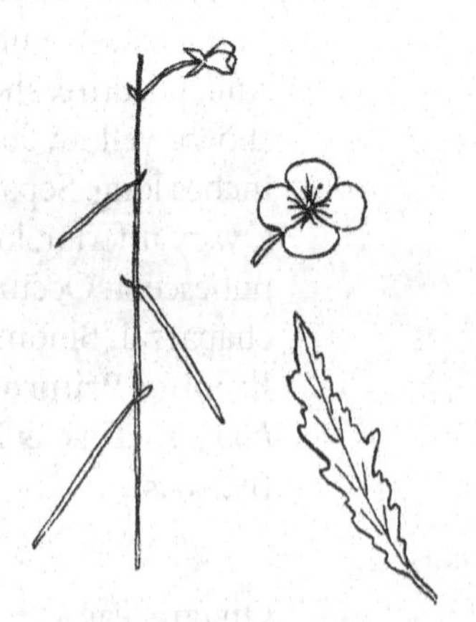

Onagraceae
Eulobus californicus Torr. & A. Gray

Mustard Evening-Primrose

---

19. Petals less than 3/16 inch long

20. Leaves linear

Branched annual, 10 to 16 inches high. Leaves linear, up to 1 inch long. Lower leaves often fascicled. Upper leaves reduced. Flowers, small, yellow. Petals about ⅛ inch long. Fruits linear, often curved, 1 to 1½ inches long; not angled. Chaparral plant, found on sandy soil. Blooms from May to June.

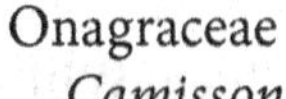

Onagraceae
Camissonia strigulosa (Fisch. & C.A. Mey.) P.H. Raven

Strigose Sun Cup

---

20. Leaves not linear

Pubescent or hairy annual, erect or with spreading branches, 2 to 20 inches high. Stems reddish, often peeling. Basal leaves petiolate, linear lanceolate. Upper leaves sessile; ½ to 1½ inches long; slightly toothed with wavy or crisped margins. Flowers small, yellow, drying orange. Petals 3/16 inch or less long. Pods linear, curved or contorted; ½ to ¾ inch long, 4-angled. Fairly common plant in dry disturbed areas. Blooms from March to May.

Onagraceae
Camissoniopsis micrantha (Spreng.) W.L. Wagner & Hoch

Small Evening-Primrose

---

16. Flowers pink, blue, violet, or white

17. Petals 2-cleft or lobed. Seeds with a tuft of hair on one end

18. Plant of moist places

Erect perennial up to 3½ feet high with soft appressed hairs on stem. Leaves 1 to 1½ inches long, glabrous, ovate to lanceolate; toothed. Flowers small with 4, pale pink to purplish, 2-cleft petals, ⅛ to 3/16 inch long. Fruit more or less reddish, 1½ to 2½ inches long. Blooms from July to September. Found in moist places.

Onagraceae
Epilobium ciliatum Raf.

Willowherb

---

18. Plant of dry, open places

Erect annual with a glabrous exfoliating stem. Leaves alternate, linear to lanceolate slightly toothed. Smaller leaves are fascicled in the axils. Petals are deeply 2-cleft; pink to white, ¼ inch long. Seeds have a tuft of hairs at one end. Found in dry open places throughout California. Blooms from June to September.

Onagraceae
Epilobium brachycarpum C. Presl

Field Willowherb

---

17. Petals not 2-cleft or lobed. Seeds without a tuft of hairs

18. Petals clawed; narrowed at the base

19. Claw without lateral lobes. Petals ⅜ to 1 inch long. Leaves linear lanceolate

Erect, glabrous annual 1 to 3½ feet high. Leaves linear lanceolate, ½ to 1 inch long. Flowers showy, nodding in the bud. Sepals 4; united and turned to one side. Petals 4; clawed; blue-violet; ¼ to ¾ inch long. Stamens 8; 4 anthers red, 4 cream. Common chaparral plant of late spring. Blooms from May to June.

Onagraceae
Clarkia unguiculata Lindl.

Elegant Clarkia

---

19. Claw with lateral lobes at base. Petals ¼ to ½ inch long. Leaves ovate

Erect annual 8 to 44 inches high. Leaves are subopposite; glabrous, ovate oblong, entire; ¾ to 3 inches long on petioles ½ to 1 inch long. Stem axis and buds nodding but becoming erect in flower. Petals clawed at the base, pink to reddish purple, ¼ to ½ inch long. Occurs on dry slopes throughout the area. Flowers from May to July.

Onagraceae
Clarkia rhomboidea Douglas

Diamond Clarkia

---

18. Petals not clawed

19. Flowers white

20. Petals 3/16 to ⅜ inch long

21. Flowers in a leafy raceme; nodding in the bud. Capsule with a very short beak. From Los Angeles county south

Erect annual 8 to 28 inches high, slightly pubescent. Leaves linear lanceolate; ½ to 1 inch long, entire or minutely toothed. Flowers small, white, nodding in the bud. Sepals 4; reflexed in pairs and turned to one side. Petals about 3/16 to ⅜ inch long. Inhabitant of shady areas in the chaparral. Blooms from March to May. Less common is **Ramona Clarkia**, *Clarkia similis* H. Lewis and W.R. Ernst, which resembles *C. epilobioides* except that petals are often flecked with purple and narrowed to a very short claw.

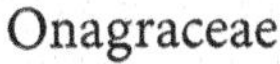

Onagraceae
Clarkia epilobioides (Torr. & A. Gray) A. Nelson & J.F. Macbr.

Canyon Clarkia

---

21. Flowers in a compact spike. Capsule curved; narrowed into a slender beak. Los Angeles County and north

Erect somewhat fleshy annual with exfoliating epidermis. Leaves mostly near the base, subentire, ¾ to 3 inches long. Flowers white, opening in the evening. Petals 4; 3/16 inch long. Flowers in a compact spike up to 12 inches long. Shredding primrose grows away from the coast, from Los Angeles County north. Fruit a coiled or contorted sessile capsule.

Onagraceae
Eremothera boothii (Douglas) W.L. Wagner & Hoch subsp. *decorticans* (Hook. & Arn.) W.L. Wagner & Hoch

Shredding Evening-Primrose

---

20. Petals ¾ to 1¼ inches long

Perennial, 4 to 20 inches high. Stems with appressed hairs and exfoliating or peeling epidermis. Leaves lanceolate, ¼ to 2¼ inches long; entire to toothed. Flowers open in the evening. Petals 4; white, turning pink with age; ¾ to 1¼ inches long. Buds nodding. Fruits ¾ to 2 inches long. In sandy, dry places. Blooms from April to June.

Onagraceae
Oenothera californica (S. Watson) S. Watson

California Evening-Primrose

---

19. Flowers lavender or pale pink to purple or dark red

20. Flowering axis and buds erect

Erect annual, 4 to 20 inches high. Leaves linear to linear lanceolate, ½ to 2 inches long. Flowers purple to deep red or maroon; axillary. Sepals 4, green; reflexed at maturity. Petals 4, lavender to purple or deep red, often with a darker mark near the center margin. Petals ¼ to ⅝ inch long. Common on grassy areas of chaparral. Blooms from April to July.

Onagraceae
Clarkia purpurea (Curtis) A. Nelson & J.F. Macbr. subsp. *quadrivulnera* (Lindl.) H. Lewis & M. Lewis

Purple Clarkia

---

20. Flowering axis or buds nodding

21. Petals ½ to 1½ inches long. North of San Diego County

22. Flowering axis erect; buds nodding

Glabrous annual 1 to 3 feet high. Leaves linear to lanceolate, 1 to 3 inches long. Buds nodding. Sepals ⅜ to ¾ inch long, reddish purple; united and turned to one side. Petals showy, fan-shaped; ⅜ to 1¼ inches long, pink or lavender at the tips, white toward base and usually flecked with violet marks; dark at base. Pod 1 to 1½ inches long. Common on dry slopes of chaparral in late spring. Orange and Riverside Counties north to Monterey.

Onagraceae
Clarkia bottae (Spach) H. Lewis & M. Lewis

Punchbowl Godetia

---

22. Flowering axis curved in bud; erect in maturity

Erect annual, 8 to 20 inches tall, finely pubescent above. Leaves alternate, lanceolate to lance-linear, ¾ to 2 inches long and ⅛ to ¼ inch across. Flowering axis curved in bud but erect at maturity. Buds nodding. Flowers with a purple to lavender margin and a red-purple base with a whitish center area, often flecked with red-purple. Petals 4; ⅜ to 1½ inches long. Fruit a capsule ¾ to 2 inches long; beaked. Grows on dry slopes in the chaparral from Los Angeles County northward. Blooms from April to July.

Onagraceae
Clarkia cylindrica (Jeps.) H. Lewis & M. Lewis

Speckled Clarkia

---

21. Petals less than ½ inch long. San Diego County

Erect glabrous annual, 8 to 28 inches high. Leaves ovatelanceolate; lower leaves toothed but upper ones entire. Flowering axis straight but buds deflexed or nodding. Petals pink to rose, ⅜ to ½ inch long. Sepals united and turned to one side. Found on dry slopes; San Diego County. Flowers from May to June.

Onagraceae
Clarkia delicata (Abrams) A. Nelson & J.F. Macbr.

Delicate Clarkia

---

10. Petals 3. Sepals 2

See *Calyptridium*, page 56.

5. Petals fused, at least at the base **Boraginaceae—Borage Family**

6. Sepals with an extra reflexed appendage

---

7. Straggling, climbing plant with retrorse prickles. Corolla lavender purple

Straggling, climbing annual with watery brittle stems and retrorse prickles. Leaves pinnately divided; lobes often pointed backward. Petiole winged and clasping the stem with an auricled base. Flowers lavender to purple, in terminal clusters. Sepals 5, each with a reflexed appendage; about ¼ to ⅜ inch long. Petals fused at base; 5-lobed, mostly ½ to 1 inch across. Common in brushy areas of the chaparral. Blooms from March to May.

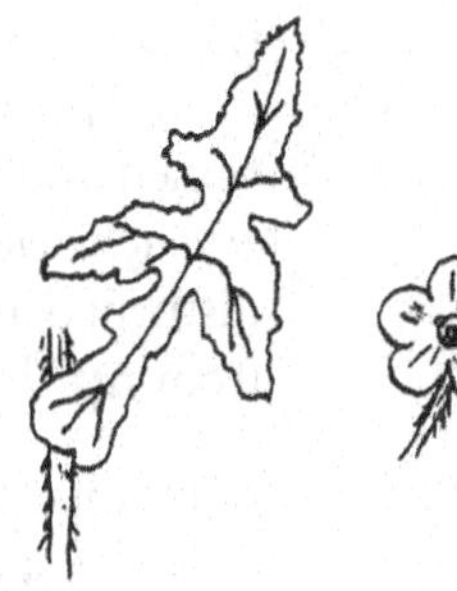

Boraginaceae
Pholistoma auritum (Lindl.) Lilja

Fiesta Flower

---

7. Plant not straggling or climbing with retrorse prickles. Corolla pale blue

Slender spreading annual with weak, watery stems 4 to 12 inches long. Leaves opposite, pinnately divided into rounded lobes; pubescent to hispid. Sepals 5; with reflexed appendages. Corolla pale blue with a lighter center, frequently speckled with purple. Corolla rotate, mostly ½ to 1 inch across, although it may be larger. Occurs in moist areas of the chaparral. Blooms from February to June.

Boraginaceae
Nemophila menziesii Hook. & Arn.

Baby Blue-Eyes

---

6. Sepals without an extra appendage
  7. Flowers in a loose or tight coiled inflorescence (Not coiled, page 106)
    8. Flowers blue, violet, or purple (Flowers other colors, page 104)
      9. Leaves simple, toothed
      10. Plant 4 to 16 inches high. Flowers dark blue to purple

Erect annual 4 to 16 inches high; hispid and glandular pubescent. Leaves ovate, ½ to 2 inches long, dentate; petiolate below but sessile above. Flowers in a loose coiled inflorescence. Calyx lobes spathulate, 3⁄16 to ¼ inch long. Corolla dark blue to purple with a pale center; rotate; ⅜ to ¾ inch long. Stamens well exserted. Common on dry slopes, particularly after burns. Blooms from March to May. **Wild Canterbury Bells,** *Phacelia minor* (Harv.) F. Zimm., is similar to *P. parryi* except that flowers are more tubular and ½ to 1½ inches long.

Boraginaceae
Phacelia parryi Torr.

Parry's Phacelia

---

10. Plant 20 to 40 inches high. Flowers pale violet

Coarse, erect annual 20 to 40 inches high; hispid and glandular pubescent. Leaves ovate, dentate, 2 to 6 inches long. Flowers in a dense cluster; showy. Calyx lobes ovate; about 3⁄16 to ¼ inch long. Corolla showy, pale blue to violet with lighter center, often veined and speckled with dark purple. Flowers ½ to 1¼ inches long; stamens well exserted. Common in disturbed areas of chaparral, such as burns. Blooms from April to June.

Boraginaceae
Phacelia grandiflora (Benth.) A. Gray

Large-Flowered Phacelia

---

9. Leaves pinnatifid dissected
10. Plant perennial

Somewhat coarse perennial, 1 to 5 feet high, with a woody root crown. Stems glandular pubescent with some longer hairs. Leaves pinnate, the lobes toothed or pinnatifid. Flowers in a tight coiled inflorescence. Calyx lobes spathulate, ¼ inch long. Corolla dirty white to pale blue, ¼ to ⅜ inch long. Occurs in chaparral areas near coast. Blooms from April to August.

Boraginaceae
Phacelia ramosissima Lehm.

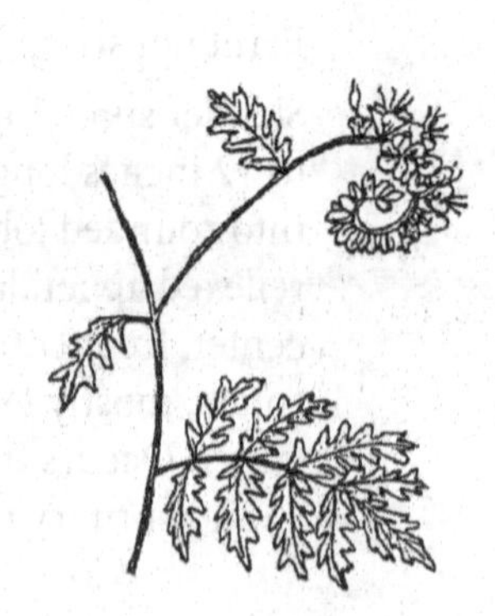

Branching Phacelia

---

10. Plant annual

11. Plant hispid. Hairs ⅛ inch long

Hispid annual, 8 to 24 inches high. Leaves pinnately compound; the lobes toothed; oblong. Flowers in a dense, coiled inflorescence, densely hispid. Calyx lobes linear to spathulate hispid. Corolla dirty white to pale lavender; ¼ to ½ inch long. Calyx lobes becoming ⅜ inch in fruit. Very abundant on dry slopes of the chaparral. Blooms from March to May.

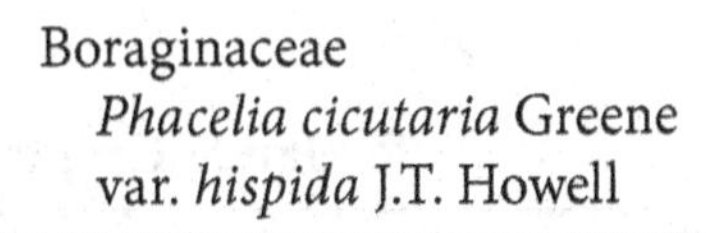

Boraginaceae
Phacelia cicutaria Greene
var. *hispida* J.T. Howell

Caterpillar Phacelia

---

11. Plant pubescent. Hairs less than ⅛ inch long

12. Calyx lobes ovate to lanceolate

Erect, mostly branched annual 8 to 32 inches high. Herbage pubescent. Leaves pinnatifid into lanceolate divisions which are deeply toothed or lobed. Calyx lobes ovate to lanceolate, somewhat unequal. Corolla blue, about ¼ inch long. Occurs on dry slopes of the chaparral. Also present in the desert. Blooms from March to June.

Boraginaceae
Phacelia distans Benth.

Common Phacelia

---

12. Calyx lobes linear to linear lanceolate

Erect, branched, pubescent annual up to 3 feet tall. Leaves pinnatifid into lanceolate or oblong divisions which are toothed or lobed. Calyx lobes linear to linear lanceolate, densely hispid. Corolla blue, ¼ to ⅜ inch long. Occurs on dry chaparral slopes and also in the Mojave Desert. Blooms from March to May.

Boraginaceae
Phacelia tanacetifolia Benth.

Tansy Phacelia

---

8. Flowers white, cream, yellow, orange, or red

9. Flowers white

10. Plant hirsute or hispid

11. Stem and roots without purple dye

Stiff, erect, bristly annual 4 to 24 inches high. Leaves linear to lanceolate, ½ to 2 inches long. Flowers in a tightly coiled (scorpioid) spike, 2 to 6 inches long. Calyx 5-lobed, oblong, 3/16 inch long; bristly hispid. Corolla white, rotate; ⅛ to ¼ inch across. Common on dry chaparral slopes. Blooms from March to July.

Boraginaceae
Cryptantha clevelandii Greene

Cleveland's Cryptantha

---

11. Stem and roots with purple dye

12. Plant erect. Flowers ¼ to 5/16 inch across. Calyx enclosing fruit

Erect, hispid annual, 8 to 20 inches high. Basal leaves in a rosette; oblanceolate; 1 to 4 inches long. Cauline leaves few; reduced upward; lanceolate to linear. Stem, root, and leaves with much purple dye. Flowers white; about ¼ inch across. Calyx tawny, silky villous; lobes coming together over fruit. Fruit of 1 to 2 nutlets. Common in grassy fields and on hillsides. Blooms March to May.

Boraginaceae
Plagiobothrys nothofulvus (A. Gray) A. Gray

Rusty Popcornflower

---

12. Plant prostrate. Flowers 3/16 inch across; subtended by leafy bracts. Calyx not enclosing fruits

Prostrate or decumbent plant with hispid or villous stems containing purple dye. Stems 4 to 24 inches long. Leaves alternate; linear to oblanceolate; ½ to 2 inches long. Flowers small, white; about 3/16 inch across. Calyx villous hirsute. Fruit of 4 nutlets. Blooms from March to May.

Boraginaceae
Plagiobothrys canescens Benth.

Valley Popcornflower

---

10. Plant glaucous

Glaucous, somewhat fleshy perennial with spreading stems. Leaves succulent, spathulate, ¼ to 1½ inches long. Flowers in a tightly coiled inflorescence ¾ to 4 inches long. Calyx lobes ¹⁄₁₆ inch long. Corolla white; often lavender towards the center; ⅛ to ¼ inch across. Common on alkaline or saline soils. Frequent inhabitant along the beach.

Boraginaceae
Heliotropium curassavicum L.

Seaside Heliotrope

---

9. Flowers pale yellow to orange or red

10. Plants not succulent. Leaves not basal

11. Flowers orange-yellow. Leaves simple, entire. Herbage hirsute

Bristly annual 8 to 32 inches high. Leaves linear to lanceolate, ¾ to 6 inches long. Flowers in a tightly coiled inflorescence 2 to 8 inches long. Calyx 5-lobed; ¼ to ⅜ inch long; bristly. Corolla orange-yellow; salverform; ⅛ to ¼ inch across; ¼ to ⅜ inch long. Very common in open grassy places. Blooms March to June.

Boraginaceae
Amsinckia intermedia Fisch. & C.A. Mey.

Common Fiddleneck

---

11. Flowers pale yellow; bell-shaped. Leaves pinnatifid. Herbage pubescent

Glandular pubescent annual 4 to 20 inches high. Leaves pinnatifid into oblong divisions. Flowers in a loosely coiled inflorescence, pendulous and bell-shaped. Calyx lobes lanceolate-ovate, ¼ to ⅜ inch long. Corolla bell-shaped; creamy yellow; ¼ to ½ inch long. Inhabitant of dry areas of the chaparral. Common after burns. Blooms from April to July.

Boraginaceae
Emmenanthe penduliflora Benth.

Whispering Bells

---

10. Plants succulent. Leaves basal

See live-forever, *Dudleya*, pages 78 to 81.

7. Flowers not in a coiled inflorescence

8. Leaves sharp needle-pointed; fascicled. Low shrub

Low branched shrub 1 to 3½ feet high. Leaves sharp needle-pointed. Linear; fascicled; ⅛ to ½ inch long. Flowers showy in terminal clusters; pink, lavender, or white. Calyx 5-cleft. Corolla ¾ to 1 inch long; salverform, 5-lobed; lobes about ½ inch long. Common shrub on dry sandy slopes. Blooms from March to June.

Polemoniaceae
Linanthus californicus (Hook. & Arn.) J.M. Porter & L.A. Johnson

Prickly Phlox

---

8. Leaves not sharp needle-pointed

9. Flowers large, showy. Petals more than 1 inch long

10. Leaves kidney-shaped. Plant found only on dunes along the coast

Prostrate, somewhat fleshy perennial with glabrous stems 4 to 20 inches long. Leaves kidney-shaped; ¾ to 2 inches wide; not as long as wide. Flowers large, showy, pale pink to purplish. Sepals ovate, ½ to ¾ inch long. Corolla funnelform; 1½ to 2¼ inches long. Common plant in sand along the coast. Blooms from April to August.

Convolvulaceae
Calystegia soldanella (L.) R. Br.

Beach Morning-Glory

---

10. Leaves not kidney-shaped. Plant not found only on coastal dunes

11. Plants vine-like with tendrils—See **SECTION IV**, page 179.

11. Plants not vine-like; without tendrils

12. Flowers more than 4 inches long

Coarse perennial, somewhat spreading; stems 2 to 4 feet long. Herbage gray pubescent. Leaves ovate, unequal at base; 1½ to 4½ inches long. Flowers large, showy, pale lavender to dirty white. Calyx tubular, 2½ to 4 inches long. Corolla long funnelform, 6 to 8 inches long, violet-tinged. Fruit a prickly capsule with spines ¼ to ½ inch long. Seeds large, **POISONOUS**. Common in dry sandy areas. Blooms from April to October.

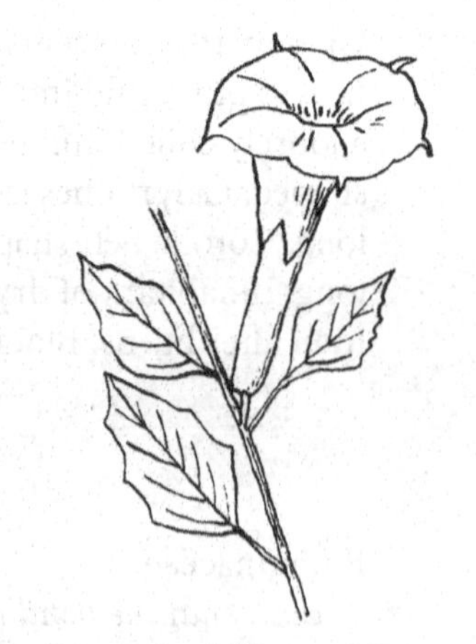

Solanaceae
Datura wrightii Regel

Jimson Weed "Sacred Datura"

---

12. Flowers less than 4 inches long

Glandular pubescent annual; erect; 1 to 3 feet high. Leaves ovate to lanceolate, 2 to 8 inches long. Flowers in terminal racemes. Calyx teeth linear lanceolate. Corolla salverform, white, 1½ to 2½ inches long. Common in valleys and dry stream beds. Blooms from May to October.

Solanaceae
Nicotiana quadrivalvis Pursh

Indian Tobacco

---

9. Flowers smaller. Petals less than 1 inch long
10. Stamens exserted; united or forming a tube
11. Corolla white; deeply 5-lobed
12. Corolla lobes more than ⅛ inch long

Slightly woody perennial 3 to 6 feet high with angled stems and sparingly pubescent herbage. Leaves ovate, sinuate dentate, ¾ to 4 inches long. Flowers small, in umbels. Peduncles ¼ to 1 inch long. Pedicels ⅛ to ½ inch long. Corolla rotate, deeply 5-parted; ⅜ to ¾ inch across. Stamens 5; anthers forming a tube which projects forward. Fruit a black berry. Common plant of the chaparral. Blooms most of the year. The berries are **POISONOUS** and should not be eaten.

Solanaceae
Solanum douglasii Dunal

White Nightshade

---

12. Corolla lobes less than ⅛ inch long

Annual or perennial with straggling stems 1 to 2 feet long. Leaves ovate, sinuate dentate, 1½ to 4 inches long. Flowers similar to above, except smaller. Corolla 3/16 to ¼ inch across. Fruit a black berry. Common weed in waste places and fields. Blooms from April to November. The berries are **POISONOUS** and should not be eaten.

Solanaceae
Solanum americanum Mill.

Small-Flowered Nightshade

---

11. Corolla blue, lavender, or red-violet

12. Corolla blue, shallowly 5-lobed

Glandular pubescent perennial, slightly shrubby; 1½ to 3 feet tall. Leaves ovate with wavy margins or slightly toothed; ¾ to 1½ inches long. Pedicels usually longer than the peduncles, averaging ¾ inch long. Flowers in umbels on peduncles ⅛ to ½ inch long. Calyx shallowly 5-lobed; about ⅛ inch long. Corolla bright blue to blue-violet; ½ to 1 inch across; very shallow lobed and appearing 5-angled. Stamens 5; anthers formed into a tube. Fruit a greenish berry about ¼ inch or slightly more across. Common plant on dry chaparral slopes. Blooms from February to June. Berries should not be eaten because they are considered **POISONOUS.**

Solanaceae

Solanum xanti A. Gray

Purple Nightshade

---

12. Corolla lavender to red-violet; deeply 5-lobed

See *Dodecatheon*, page 77.

10. Stamens not exserted

11. Plant succulent

See live-forevers, *Dudleya*, pages 78 to 81.

11. Plant not succulent

12. Leaves opposite or whorled (Leaves alternate, page 110)

13. Leaves whorled or appearing so

14. Flowers in a dense head. Corolla 5-lobed, with a very long, filiform tube. Plant erect

Slender erect annual, 2 to 12 inches high; pubescent. Leaves parted into 5 to 9 filiform segments and thus appearing whorled. Flowers terminal in a dense head subtended by leaflike bracts. Calyx deeply 5-parted with linear lobes, stiff, with cilia; becoming almost prickly in fruit. Corolla salverform with a very narrow tube; white to pale lilac; yellowish at the base. Tube 2 to 3 times the length of the calyx. Corolla about $^3/_{16}$ to ¼ inch across. Occurs in grassy areas of the chaparral. Blooms from April to May.

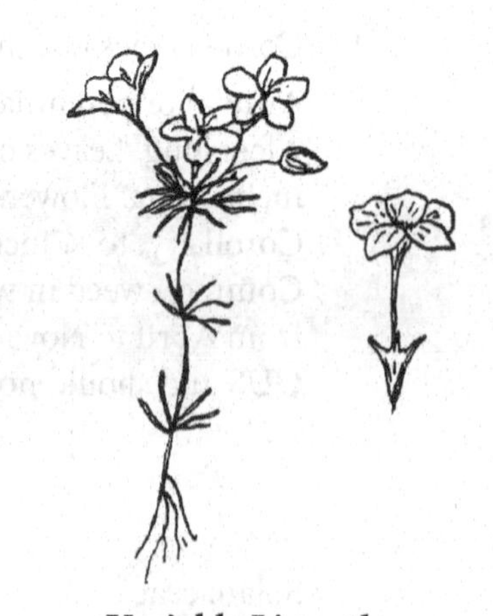

Polemoniaceae

Leptosiphon parviflorus Benth.

Variable Linanthus

---

14. Flowers not in a dense head. Corolla 4-lobed. Plant straggling
    See *Galium*, page 59.

13. Leaves opposite

14. Plant low, spreading. Flowers salmon-colored. Common weed
    See *Anagallis*, page 85.

14. Plant erect. Flowers not salmon-colored

15. Plant low, 2 to 4½ inches high. Leaves filiform, entire

Low annual with slender erect stems 2 to 4½ inches high. Leaves filiform, opposite, entire. Flowers large showy, mostly solitary. Calyx deeply 5-lobed; lobes linear, ⅜ to ⅝ inch long. Corolla rotate or funnelform, pink to lilac, yellow below with dark spots at the base. Corolla ⅜ to 1 inch long. Common spring annual on open rocky slopes. Blooms from February to April.

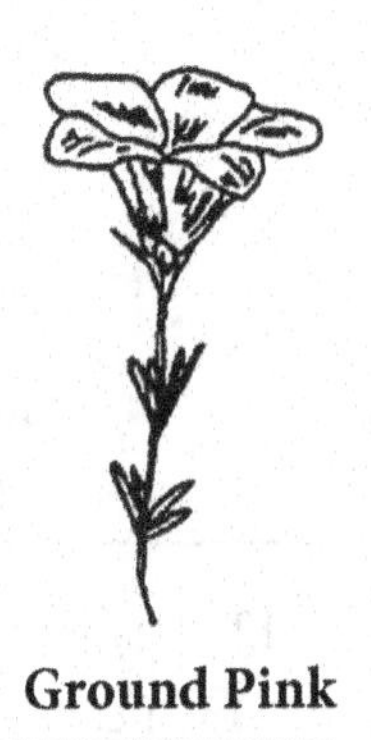

Polemoniaceae
Linanthus dianthiflorus (Benth.) Greene

Ground Pink

---

15. Plant taller, 4 to 20 inches high

16. Leaves simple, lanceolate to oblong-ovate. Flowers blue or rose

17. Herbage with distinct mint odor. Flowers in a terminal head
    See *Monardella*, page 128.

17. Herbage not mint odored. Flowers not in a terminal head

18. Leaves entire. Flowers rose, ⅜ to 1 inch long

Glabrous annual 4 to 12 inches high with opposite leaves. Leaves sessile; ovate to oblong; ⅜ to 1 inch long. Corolla salverform; 5-lobed. Lobes rose-purple, up to ½ inch long and throat white. Calyx 5-lobed. Stamens 5. Fruit a capsule. Common on dry slopes. Blooms from March to July.

Gentianaceae
Zeltnera venusta (A. Gray) G. Mans.

California Centaury

---

18. Leaves toothed. Flowers purple to blue, ⅛ to ⅜ inch long

19. Leaves minutely toothed. Flowers in axillary racemes, pale blue-violet
See *Veronica*, page 129.

19. Leaves coarsely toothed. Flowers in terminal spike, blue-purple
See *Verbena*, page 134.

16. Leaves pinnatifid dissected. Flowers white

Erect, hirsute annual 8 to 20 inches high. Leaves pinnatifid dissected into rounded lobes. Lower leaves opposite; upper alternate. Flowers small, white, in terminal clusters. Calyx 5-lobed. Corolla campanulate, $^3/_{16}$ to ¼ inch across. Common in shaded and disturbed areas of the chaparral. Blooms March to June. Handle with care may cause **CONTACT DERMATITIS.**

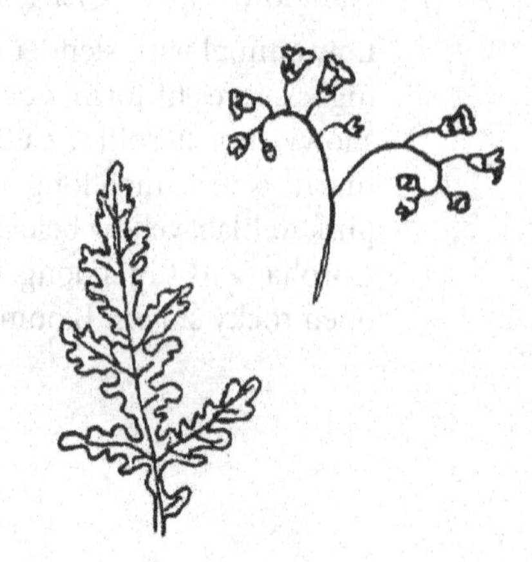

Boraginaceae
Eucrypta chrysanthemifolia (Benth.) Greene **Common Eucrypta**

---

12. Leaves alternate

13. Leaves pinnately dissected, at least below

14. Flowers in a dense, many-flowered, terminal head

15. Leaves prickly toothed. Flowering heads subtended by prickly bracts

Erect annual 2 to 8 inches high with glandular, hirsute, rigid stems. Leaves alternate, sessile, ¾ to 2 inches long and pinnately dissected with prickly teeth. Flowers are white to blue or purple in spiny-bracted heads. The bracts are ovate with spiny teeth. Flowers are narrow funnelform, 5-lobed, about ⅜ inch long. *Navarretia* grows in dry, sandy soil below 2,000 feet. It blooms from May to July. Also present in dry rocky areas is **Hooked Navarretia,** *Navarretia hamata* Greene, which is similar *N. atractyloides* except that it has a distinct skunk-like odor, and the bract tips are hooked. Blooms from April to June.

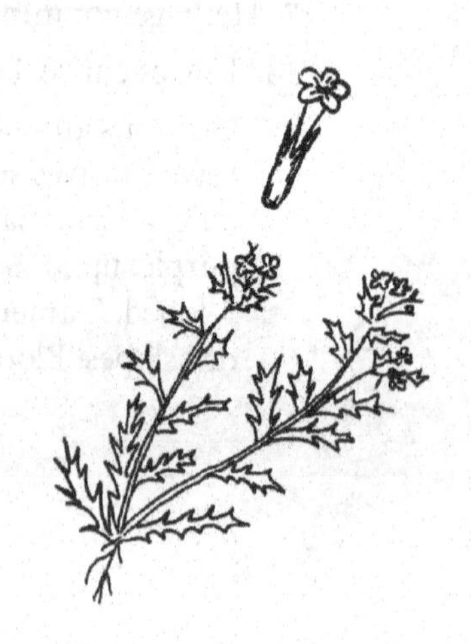

Polemoniaceae
Navarretia atractyloides (Benth.) Hook. & Arn. **Hollyleaf Navarretia**

---

15. Leaves not prickly toothed. Flowering heads not subtended by prickly bracts

Slender annual, ¾ to 3 feet high. Leaves alternate, pinnately dissected. Flowers in a dense-flowered terminal head of 10 to 25 flowers. Head ½ to 1¼ inches thick. Individual flowers pale blue, ¼ to ½ inch long. Calyx 5-lobed, 3/16 inch long. Occasional in disturbed or burned places of the chaparral. Blooms from May to July.

Polemoniaceae
Gilia capitata Sims
subsp. *abrotanifolia* (Greene) V.E. Grant

Bluehead Gilia

---

14. Flowers not in a dense, many-flowered terminal head

15. Flowers white

See *Eucrypta*, page 110.

15. Flowers pale blue to dark blue-violet

16. Flowers pale blue with yellow throat

Low, slender annual 5 to 28 inches high. Leaves pinnately divided below into linear divisions. Upper leaves simple, linear. Flowers in terminal clusters. Calyx ⅛ inch long; 5-lobed. Corolla campanulate, about ¼ inch long; lobes pale blue or white; yellow below. Common in the spring in sandy soil on grassy slopes. Blooms from March to May. Sometimes occurs in masses becoming quite conspicuous.

Polemoniaceae
Gilia angelensis V.E. Grant

Chaparral Gila

---

16. Flowers dark blue-violet

Erect annual 4 to 16 inches high with alternate leaves. Leaves pinnately lobed into linear divisions. Corolla dark blueviolet, ¼ to ⅜ inch long; funnelform with 5 lobes. Stamens 5. Flowers in a 4-flowered cluster. Fruit a subspherical capsule. Occurs on open, dry slopes. Blooms from April to June.

Polemoniaceae
Allophyllum gilioides (Benth.)
A.D. Grant & V.E. Grant

Dense False Gilia

---

13. Leaves not pinnately dissected

Heavy-scented annual with sticky glandular, pubescent stems 8 to 24 inches high. Leaves ovate to lanceolate, 1¼ to 2½ inches long, sessile or on short petioles. Flowers in panicles. Corolla is tubular to salverform, 5-lobed, ⅝ to ¾ inch long, white, tinged with violet. Cleveland's tobacco grows in sandy places along the beach or near the coast. Blooms from March to June.

Solanaceae
Nicotiana clevelandii A. Gray

Cleveland's Tobacco

---

4. Flowers irregular (with bilateral symmetry), or with sepals or petals spurred
5. Flowers papilionaceous. Leaves compound into 3 to 39 leaflets **Fabaceae—Pea Family**
(Flowers not papilionaceous. Leaves simple to pinnatifid dissected, page 121)

Papilionaceous Flower

---

6. Leaves 3-foliate
(Leaves not 3-foliate, page 116)
7. Flowers yellow
8. Plant shrubby, woody at the base
See *Acmispon glaber*, page 9
8. Plant not shrubby
9. Plant low, spreading. Common weed on lawns
10. Flowers small, ¼ inch long. Fruit a coiled bur

Glabrous, flat lying or prostrate annual. Leaves 3-foliate with laciniate stipules. Leaflets obovate, ¼ to ¾ inch long, minutely toothed. Peduncles arising from leaf axils; 2 to 5 flowered. Flowers yellow, less than ¼ inch long. Fruit a coiled pod about ¼ inch across, bearing many hooked spines. Common weed in lawns and waste places. Blooms from March to June.

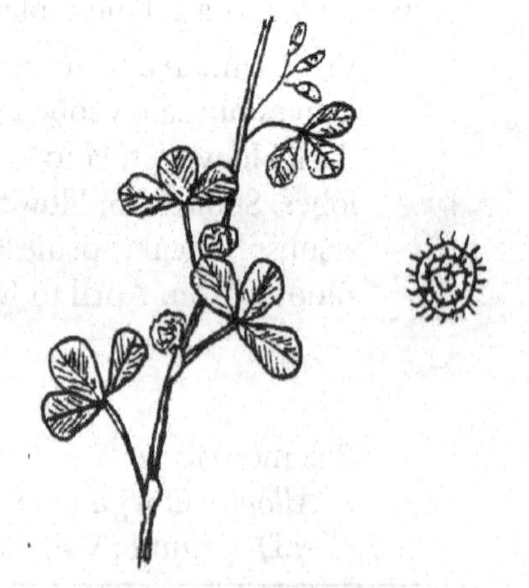

Fabaceae
Medicago polymorpha L.

California Burclover

---

10. Flowers larger, showy; ⅜ to ½ inch long. Fruit a straight pod

Spreading perennial 4 to 40 inches long. Leaves 3-foliate with leaflike stipules and almost appearing 5-foliate. Leaflets obovate, ¼ to ⅝ inch long. Flowers in 3 to 6 flowered clusters, on peduncles 2 to 5 inches long. Corolla yellow, ⅜ to ½ inch long. Pods linear, straight, ¾ to 1 inch long. Common weed of lawn and roadside. Blooms from June to September.

Fabaceae
Lotus corniculatus L.

Bird's-Foot Trefoil

---

9. Plant erect, 8 to 32 inches high. Weed of waste places

Glabrous plant 8 to 32 inches tall. Leaves 3-foliate with oblanceolate leaflets ⅜ to 1¼ inches long. Leaflets dentate with an obtuse apex. Stipules present, ¼ inch long, narrow, sharp pointed. Flowers in a spike-like, terminal raceme ¾ to 4 inches long. Corolla yellow, ⅛ inch long. Common in waste places. Blooms April to October.

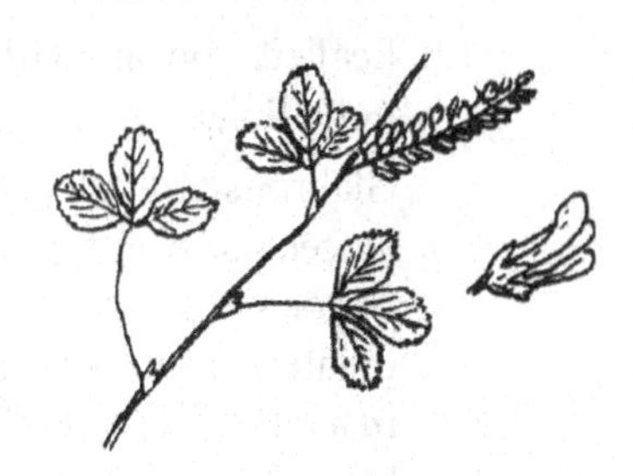

Fabaceae
Melilotus indicus (L.) All.

Sourclover

---

7. Flowers not yellow
8. Flowers in a many-flowered terminal head
(Flowers not in a terminal head, page 115)

Clovers

9. Corolla white

Glabrous perennial with runners rooting at the nodes. Leaves trifoliate, with lanceolate stipules. Leaflets obovate, emarginate and minutely toothed. Flowers white, in a terminal head. Pedicels about ⅛ inch long, becoming reflexed in age. Corolla ¼ to ⅜ inch long, white to pinkish. Frequently planted in lawns and often occurring as an escape. Blooms April to December.

Fabaceae
Trifolium repens L.

White Clover

---

9. Corolla pink to red-purple

10. Floral heads subtended by a green laciniate involucre. Flowers red-purple, ½ to ⅝ inch long. Leaflets linear

Glabrous annual 4 to 16 inches tall; erect or spreading along the ground. Leaves with prominent, ovate to lanceolate laciniate stipules. Leaves trifoliate. Leaflets linear to narrow oblong, ½ to 1½ inches long; minutely toothed. Flowering head subtended by a flat, unevenly laciniate involucre ⅜ to ⅝ inch across. Flowers papilionaceous, red-purple; paler toward the tips; ½ to ⅝ inch long. Quite common in grassy areas. Blooms from March to June.

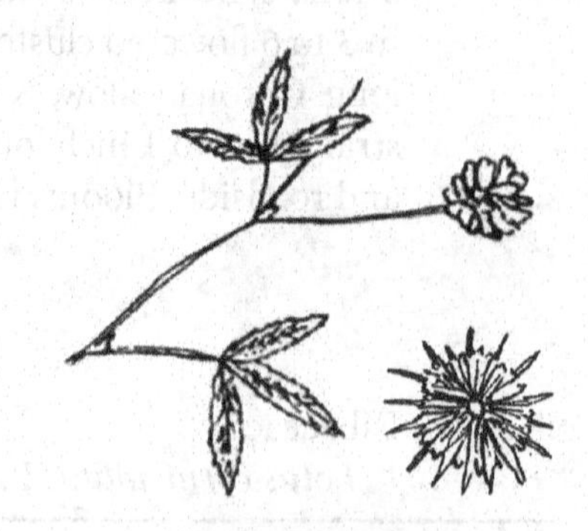

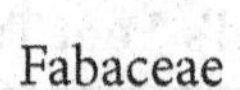

Fabaceae
Trifolium willdenovii Spreng.

Tomcat Clover

---

10. Floral head not subtended by a green involucre

11. Annual. Corolla ¼ inch long, on short pedicels which are reflexed in age. Heads not subtended by leaves

12. Leaflets obovate with emarginate apex. Calyx lobes not ciliolate

Glabrous annual 4 to 16 inches high. Leaves with lanceolate stipules ⅜ inch long. Leaflets 3-foliate on petioles ¾ to 2¾ inches long. Leaflets obovate, emarginate, minutely toothed; ¼ to ⅝ inch long. Flowers in a dense, terminal head, not subtended by an involucre. Pedicels ⅛ inch or less, becoming reflexed in age. Corolla pink to red-purple, ¼ inch long. Fairly common in grassy areas. Blooms from April to June.

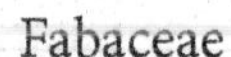

Fabaceae
Trifolium gracilentum Torr. & A. Gray

Pinpoint Clover

---

12. Leaflets obovate to oblong with obtuse apex. Calyx lobes ciliolate

Glabrous annual 8 to 20 inches high. Leaves with sharp-pointed lanceolate stipules ½ to 1¼ inches long. Leaves 3-foliate. Leaflets obovate to oblong; obtuse margined, minutely toothed. Flowers in a dense, terminal head on pedicels less than 1/16 inch long and becoming reflexed in age. Calyx ciliolate on the lobes. Corolla pink to purple, ¼ inch long. Fairly common in open grassy areas. Blooms from March to June.

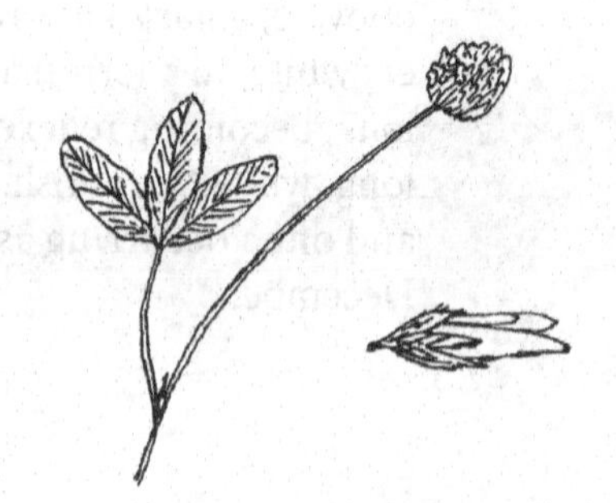

Fabaceae
Trifolium ciliolatum Benth.

Foothill Clover

---

11. Perennial. Corolla ⅜ to ¾ inch long; sessile. Heads subtended by 1 to 2 leaves

Perennial, 8 to 24 inches high with pubescent stems and leaves. Leaves 3-foliate with ovate to obovate leaflets which are ¾ to 2 inches long, entire or toothed, and often with a pale blotch in the center. Flowers in a compact head, subtended by 1 to 2 leaves. Heads ¾ to 1¼ inches across. Flowers many, sessile. Corolla papilionaceous, red or pink, ⅜ to ¾ inch long. Red clover is widely cultivated and is commonly found as an escape. Blooms from April to October.

Fabaceae
Trifolium pratense L.

Red Clover

---

8. Flowers not in a terminal head
9. Leaflets toothed. Flowers in a terminal raceme
10. Flowers violet to blue, ⅜ to ½ inch long

Smooth perennial 1 to 3 feet tall. Leaves 3-foliate with leaflike stipules ¼ inch long. Leaflets oblanceolate, dentate, ⅜ to 1 inch long; apex truncate. Flowers violet to blue (may fade to white), in a short, dense raceme. Corolla ⅜ to ½ inch long. Common plant in waste places and fields. Blooms from April to October.

Fabaceae
Medicago sativa L.

Alfalfa

---

10. Flowers white, ¼ inch long

Tall, glabrous plant 3 to 6 feet tall. Leaves trifoliate with linear, bract-like stipules. Leaflets ovate to lanceolate, toothed. Flowers white, in spike-like racemes 2 to 4 inches long. Common weed of waste places, particularly when damp. Blooms from May to September.

Fabaceae
Melilotus albus Medik.

White Sweetclover

---

9. Leaflets entire. Flowers pink-white, occurring singly; subtended by a leaflike bract

Slender, glabrous annual 6 to 32 inches high. Leaves 3-foliate with a leaflike bract at base of flower and almost appearing as an extra leaflet. Leaflets ovate-lanceolate, entire. Flowers on peduncles ⅜ to ⅝ inch long; 1-flowered and subtended by a leaflike bract. Corolla whitish with a pinkish tinge; averaging ¼ inch long. Inhabitant of dry or disturbed areas of the chaparral. Blooms from May to October.

Fabaceae
Acmispon americanus (Nutt.) Rydb. var. *americanus*

Spanish Clover

---

6. Leaves not 3-foliate

7. Leaves palmate compound

(Leaves pinnately compound, page 118)

Lupines

8. Leaflets broad, ½ to 1 inch wide

9. Plant glabrous or sparsely hairy

Erect perennial, 1 to 4 feet high, glabrous or with a few hairs. Leaves palmate compound with 7 to 9 leaflets. Leaflets broadly oblanceolate, ½ to 4 inches long. Flowers in racemes 6 to 18 inches long. Corolla blue or purplish. Found in protected canyons and in wooded areas. Blooms from April to July.

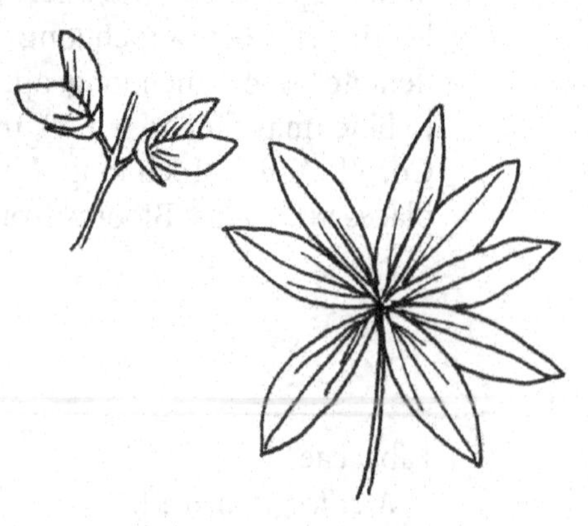

Fabaceae
Lupinus latifolius J. Agardh

Broad-Leafed Lupine

---

9. Plant hirsute with stiff stinging hairs

Robust annual covered with long, stiff, yellowish, stinging hairs. Plant 8 to 32 inches high. Leaves 5 to 8 foliate. Leaflets obovate, ½ inch across and ¾ to 2 inches long. Flowers red-violet or magenta in a terminal raceme. Corolla about ½ inch long. Inhabitant of wooded, brushy places of the chaparral. Blooms from March to May.

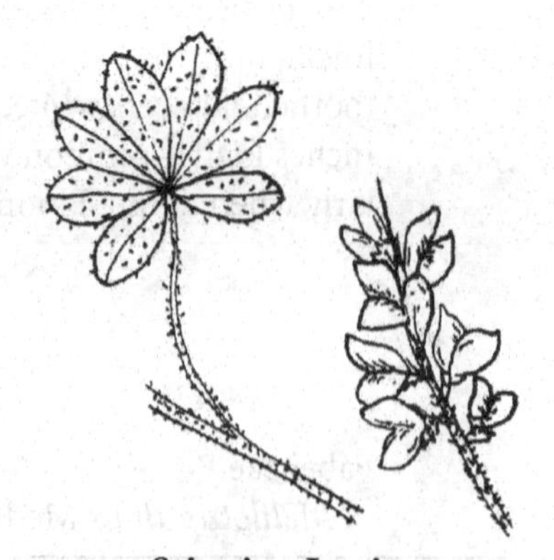

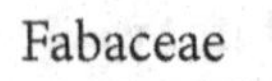

Fabaceae
Lupinus hirsutissimus Benth.

Stinging Lupine

---

8. Leaflets narrower; less than ½ inch wide

9. Plant glabrous or sparsely pubescent

10. Leaflets linear with truncate apex. Flowers not whorled

Glabrous annual 8 to 20 inches tall with dark green foliage. Leaves with 5 to 7 leaflets. Leaflets linear, ¾ to 1½ inches long, with a truncate apex. Flowers not crowded, in a terminal raceme. Corolla purple-blue, ⅜ to ½ inch long. Occurs in openings in the chaparral. Often found following burns. Blooms March to May.

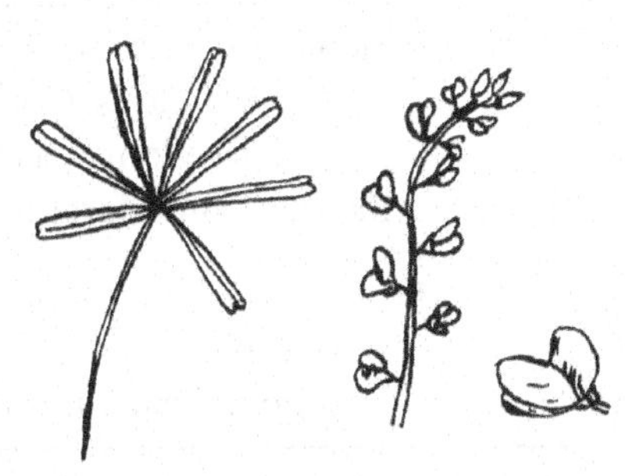

Fabaceae
Lupinus truncatus Nutt.

Collar Lupine

---

10. Leaflets obovate with rounded apex. Flowers more or less whorled.

Sparsely hairy, somewhat succulent annual 8 to 24 inches high. Leaves with 7 to 9 leaflets. Leaflets obovate, glabrous above, ¾ to 2¾ inches long. Flowers in a terminal raceme, 3 to 8 inches long. Corolla purple-blue, mostly ½ to ¾ inch long. Common in clay soil and grassy fields. One of the most abundant lupines in Ventura County. Blooms February to May and sometimes longer.

Fabaceae
Lupinus succulentus K. Koch

Arroyo Lupine

---

9. Plant pubescent to densely villous

10. Raceme long, showy; 5 to 8 inches long.

Villous annual 8 to 16 inches high. Leaves 5 to 9 foliate. Leaflets linear to oblanceolate $\frac{1}{16}$ to ⅛ inch wide; apex acute. Flowers in a terminal raceme; showy, not crowded. Corolla blue-violet; slightly less than ½ inch long. Pods densely villous. Occurs in open fields and slopes. Blooms from March to May.

Fabaceae
Lupinus sparsiflorus Benth.

Coulter's Lupine

---

10. Raceme shorter; less than 4 inches long.

11. Herbage densely villous. Leaflets ovate to lanceolate. Flowers not whorled

Densely villous, much branched annual 2 to 8 inches high. Leaves 5 to 9 foliate with narrow oblanceolate to ovate leaflets; ⅜ to ¾ inch long. Flowers in a terminal raceme; sometimes hidden among the leaves. Corolla red-purple; slightly more than ¼ inch long. Occurs in dry, gravelly places. Blooms from March to May.

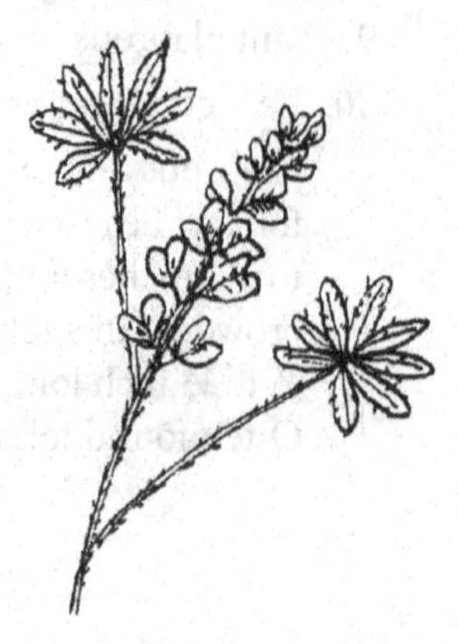

Fabaceae
Lupinus concinnus J. Agardh

Bajada Lupine

---

11. Herbage pubescent, with short hairs. Leaflets linear-oblanceolate. Flowers in definite whorls

Pubescent or villous annual 4 to 16 inches high. Leaves 5 to 7 foliate with linear to oblanceolate leaflets averaging 1⁄16 inch wide; ⅜ to 1 inch long. Flowers in a terminal raceme in definite whorls. Flowers small, blue and white; about ¼ inch long. Inhabitant of dry, gravelly places of the chaparral. Occasionally found in disturbed ground as a weed. Blooms from March to June.

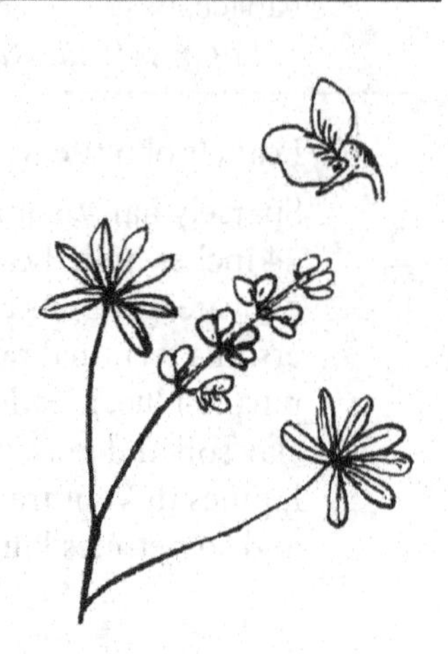

Fabaceae
Lupinus bicolor Lindl.

Miniature Lupine

---

7. Leaves pinnate compound

8. Flowers yellow; in a spike or umbel
(Flowers white to violet, page 120)

9. Leaflets 11 to 19. Flowers yellow-white, in spikes

Erect perennial, 12 to 40 inches high with sticky, pubescent stems. Leaves pinnate compound into 11 to 19 leaflets. Leaflets ovate-lanceolate to oblong, ¾ to 1¼ inches long with mucronate apex. Flowers yellowish-white; in spikes shorter than leaves. Corolla 5⁄16 to ½ inch long. Fruit an oblong pod, ½ to ⅝ inch long, bur-like, with hooked prickles. Wild licorice grows in moist ground. Blooms from May to July. Roots have a sweet flavor and may be eaten raw or used as a flavoring.

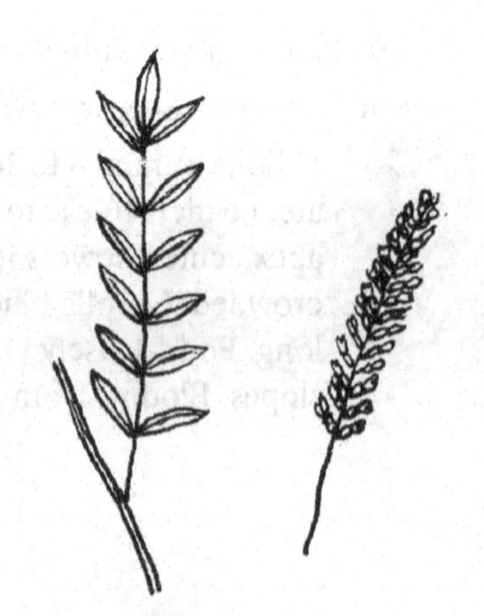

Fabaceae
Glycyrrhiza lepidota Pursh

Wild Licorice

---

9. Leaflets 5 to 9. Flowers in umbels

10. Umbels sessile. Herbage densely silvery pubescent

Prostrate perennial with silvery pubescent or tomentose herbage and 8 to 40 inch long stems. Leaflets mostly 5; oblanceolate; $\frac{3}{16}$ to ½ inch long. Flowers in sessile umbels, 3 to 8 flowered. Corolla yellow, ⅜ inch long. Pods short, silky pubescent, slightly curved and scarcely exceeding calyx. Inhabitant of dry hills and rocky places. Not too common. Blooms April to July.

Fabaceae
Acmispon argophyllus (A. Gray) Brouillet

Silver Lotus

---

10. Umbels on peduncles ⅛ to 1½ inches long

11. Corolla large; ½ to 1 inch long. Pods 1 to 1½ inches long

Somewhat straggling, pubescent perennial with stems 8 to 24 inches long. Leaflets mostly 7 to 9; obovate, mucronate, ⅜ to ¾ inch long. Flowers in umbels on long peduncles; 2 or more flowered. Flowers subtended by a leaflike bract. Corolla large, yellow, ½ to 1 inch long. Pods long, dark brown and glabrous at maturity; up to 1½ inches long. On dry chaparral slopes. Blooms from April to July.

Fabaceae
Acmispon grandiflorus (Benth.) Brouillet

Large-Flowered Lotus

---

11. Corolla smaller. Pods shorter

12. Peduncles longer than ¼ inch. Pods straight, not villous

13. Perennial. Flowers showy, $\frac{5}{16}$ to ½ inch long. Common on lawns

See *Lotus corniculatus*, page 113.

13. Annual. Flowers smaller; ¼ to ⅜ inch long

14. Leaflets oblanceolate-obovate. Pods strigose

Prostrate, mat-forming, pubescent annual, with stems 2 to 12 inches long. Leaflets 4 to 9, oblanceolate to obovate; ¼ to ½ inch long. Umbels mostly 1 to 2 flowered. Corolla yellow; ¼ to ⅜ inch long. Pods pubescent; ⅜ to slightly over 1 inch long. Inhabitant of dry, disturbed places. Blooms from March to June.

Fabaceae
Acmispon strigosus (Nutt.) Brouillet

Strigose Lotus

---

14. Leaflets obovate-round. Pods glabrous

Prostrate or decumbent annual with much branched leafy stems 4 to 12 inches long. Leaves with 5 to 8 pubescent, obovate to round leaflets. Umbels 2 to 4 flowered on peduncles ⅜ to 1½ inches long. Flowers yellow, not quite ⅜ inch long; subtended by a single bract. Legumes straight, ½ to a little over 1 inch long. Common on dry slopes and fields. Blooms from March to June.

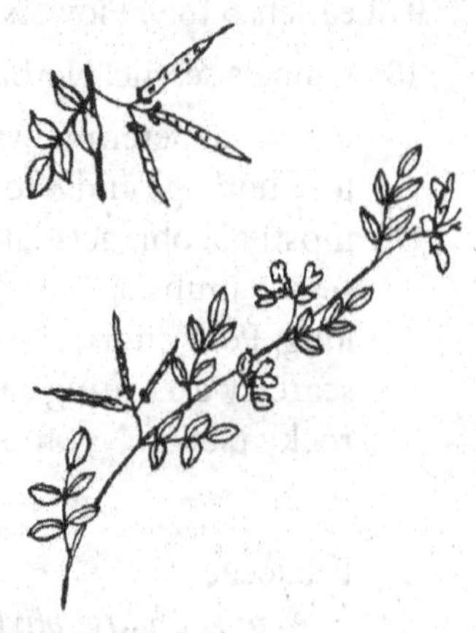

Fabaceae
Acmispon maritimus (Nutt.) D.D. Sokoloff **Coastal Lotus**

---

12. Peduncles ¼ inch long. Pods curved, villous; with a long incurved beak. Herbage villous. Corolla ⅛ to 3/16 inch long

Prostrate perennial sometimes forming mats. Stems 12 to 40 inches long; silky villous when young. Leaves pinnate compound with 4 to 6 obovate to oblanceolate leaflets. Flowers yellow; about ¼ inch long. Calyx villous. Flowers in 4 to 10 flowered umbels on peduncles ¼ inch long. Pods short; ¼ inch long; villous; curved, with a long, incurved beak. Blooms from March to October.

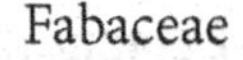
Fabaceae
Acmispon heermannii (Durand & Hilg.) Brouillet **Heermann's Lotus**

---

8. Flowers white to violet; in a terminal raceme
9. Plant without tendrils
10. Flowers greenish-white. Leaflets 15 to 39
11. Pods inflated; ovoid; pubescent

Erect perennial with hairy stems 8 to 24 inches long. Leaves pinnate compound into 15 to 39 ovate to oblanceolate leaflets up to 1 inch long. Flowers nodding; greenish white in a 12 to 36 flowered raceme. Pods ¾ to 2 inches long; inflated; ovoid; and pendulous on a stipe ¼ to ⅝ inch long. Pods papery and pubescent. Occurs in sandy areas and hills near the coast. Found from Ventura County south. Blooms February to June.

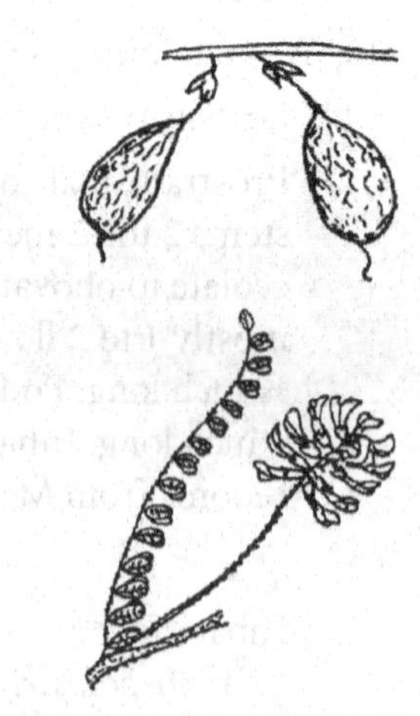

Fabaceae
Astragalus trichopodus (Nutt.) A. Gray
var. *lonchus* (M.E. Jones) Barneby **Southern California Milkvetch**

---

11. Pod not inflated; oblanceolate; glabrous

Erect perennial 1 to 3 feet high. Leaves pinnate compound with 15 to 39 leaflets. Flowers greenish white. Fruit a glabrous, compressed pod ½ to 1½ inches long; elliptic to oblanceolate. Pod not inflated. Found on dry hillsides from Los Angeles County north. Blooms February to June and sometimes in fall and winter.

Fabaceae
Astragalus trichopodus (Nutt.) A. Gray var. *phoxus* (M.E. Jones) Barneby

Antisell Milkvetch

---

10. Flowers white, tinged with violet. Leaflets 7 to 13

Somewhat straggling plant with slightly hairy stems 2 to 12 inches high. Leaves pinnate compound into 7 to 13 oblanceolate or cuneate leaflets. Leaflets glabrous above; averaging ¼ inch long. Flowers whitish, tinged with violet; occurring in 4 to 15 flowered racemes. Peduncles ½ to 2½ inches long. Fruit a round to ovate pod; compressed; and hanging downward. Flowers small. Occurs on grassy hillsides from Santa Barbara to San Diego.

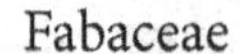

Fabaceae
Astragalus gambelianus E. Sheld.

Gambel Milkvetch

---

9. Plant with tendrils—**SECTION IV**, page 179.

5. Flowers not papilionaceous. Leaves simple to pinnatifid dissected, but not compound into separate leaflets
6. Leaves deeply lobed or pinnatifid dissected (Leaves toothed or entire, page 126)
7. Leaves palmately cleft or divided. Flowers with a spur
8. Flowers purple-blue to whitish
9. Leaves divided into linear divisions

Slender simple stem 1 to 3 feet tall. Leaves few, palmately parted or divided and then further divided into linear divisions. Leaf blades 1 to 3 inches across. Flowers in a loose, few-flowered raceme on pedicels ½ to 2½ inches long. Sepals purple-blue, ⅜ to ⅝ inch long. Spur straight, same length as sepals. Upper petals whitish or blue; lower ones dark blue. Stamens numerous. Fairly common on chaparral slopes. Blooms from April to May.

Ranunculaceae
Delphinium parryi A. Gray subsp. *parryi*

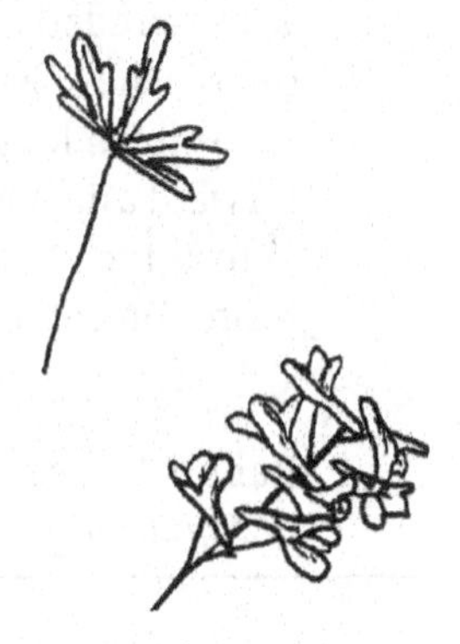

Parry's Larkspur

---

9. Leaves divided into wider divisions

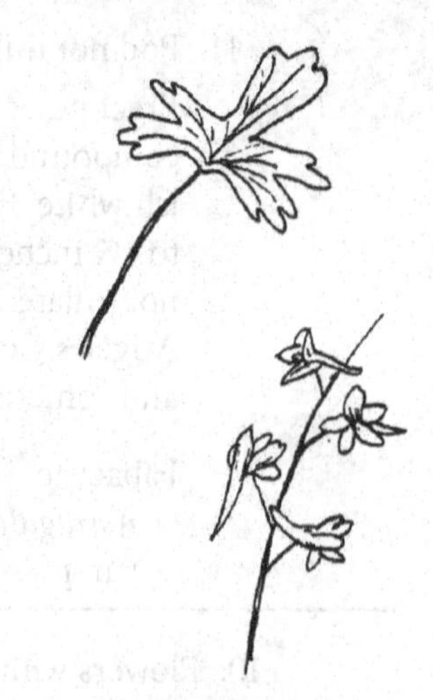

Glabrous, slender, simple-stemmed herb 8 to 20 inches tall. Leaves palmately parted in 3 lobes; the lateral divisions deeply 2-cleft. Leaves few; 1½ to 3½ inches across. Basal leaves long petioled. Flowers in a loose, few-flowered raceme on long, thin pedicels. Pedicels ½ to 3 inches long. Sepals dark blue, about ½ inch long; spur about same length. Upper petals whitish; lower ones whitish with blue veins or bluish. Inhabitant of shaded canyons of the chaparral. Blooms April to May.

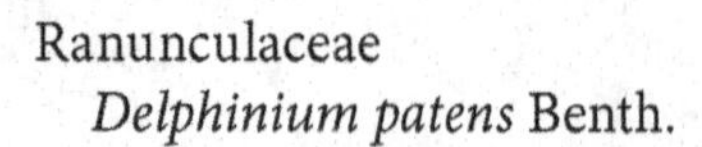

Ranunculaceae
Delphinium patens Benth. **Spreading Larkspur**

---

8. Flowers scarlet

9. Flowers in a terminal raceme. Posterior sepal formed into a long spur

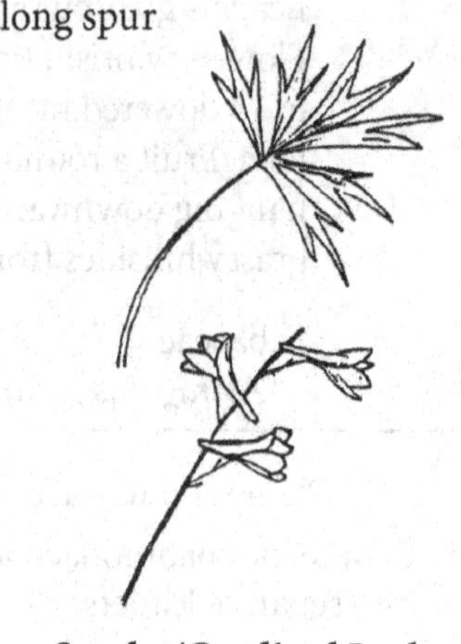

Tall, robust, thick-stemmed plant with erect, simple stems; 3 to 6 feet tall. Leaves mostly withered at time of flowering; palmately 5-parted; the divisions further divided into linear, spreading segments. Flowers in a many-flowered, terminal raceme on pedicels ¾ to 2¼ inches long. Sepals scarlet, ⅜ to ⅝ inch long; spur up to ¾ inch long. Petals yellowish with scarlet tips. Inhabitant of dry open areas in chaparral woods. Blooms in late spring from May to July.

Ranunculaceae
Delphinium cardinale Hook. **Scarlet/Cardinal Larkspur**

---

9. Flowers nodding at ends of branches. Petals each having a spur projecting backwards

Glabrous perennial with branched stems 1 to 3½ feet high. Leaves palmately cleft and then further lobed with rounded or obtuse apex. Flowers nodding at ends of branches. Sepals 5; scarlet, spreading or reflexed, ⅝ to 1 inch long. Petals 5; with a yellow projection forward and a long spur backward. Spurs scarlet, ⅜ to ¾ inch long. In moist woods along streams. More common in mountains or farther north.

Ranunculaceae
Aquilegia formosa DC. **Western Columbine**

---

7. Leaves pinnately divided or dissected

8. Leaves 2 to 3 times pinnatifid

9. Flowers heart-shaped; cream to yellow

10. Flowers bright yellow; ½ inch long

Glaucous perennial with stout, erect stems 15 to 45 inches high. Leaves pinnately dissected into narrow, linear divisions. Flowers in a dense, many-flowered panicle. Corolla bright yellow, slightly heart-shaped; ½ inch long. Upper portion forming a sac-like base. Inhabitant of disturbed areas of the chaparral, particularly after burns. Blooms from April to September.

Papaveraceae
Ehrendorferia chrysantha (Hook. & Arn.) Rylander

Golden Eardrops

---

10. Flowers cream colored; ¾ to 1 inch long

Similar to the above, except that the flowers are slightly larger and cream-colored rather than golden yellow. Petal tips tinged with purple. Also occurs in dry, disturbed areas of the chaparral. Blooms from May to July.

Papaveraceae
Ehrendorferia ochroleuca (Engelm.) Fukuhara

White Eardrops

---

9. Flowers not heart-shaped or cream to yellow

10. Flowers red in a dense terminal spike

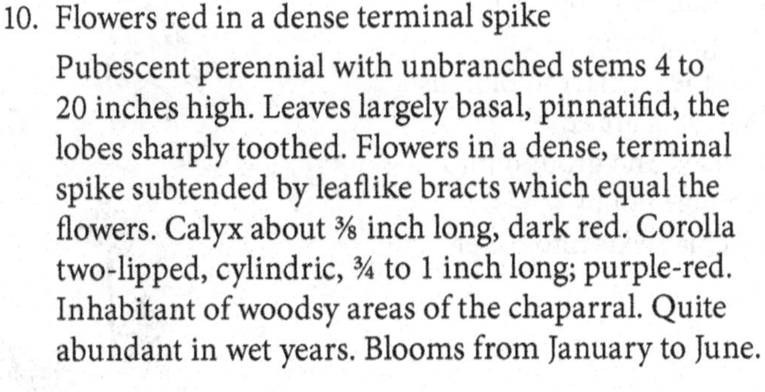

Pubescent perennial with unbranched stems 4 to 20 inches high. Leaves largely basal, pinnatifid, the lobes sharply toothed. Flowers in a dense, terminal spike subtended by leaflike bracts which equal the flowers. Calyx about ⅜ inch long, dark red. Corolla two-lipped, cylindric, ¾ to 1 inch long; purple-red. Inhabitant of woodsy areas of the chaparral. Quite abundant in wet years. Blooms from January to June.

Orobanchaceae
Pedicularis densiflora Hook.

Warrior's Plume

---

10. Flowers blue, in interrupted whorls

See *Salvia columbariae*, page 124.

8. Leaves once pinnate or pinnatifid

9. Leaves opposite or basal

10. Leaves mostly basal. Flowers blue to purple, in 1 to 3 interrupted axillary whorls. Plant of dry areas

11. Plant densely white-woolly. Leaves spinose toothed. Corolla pale lavender

Erect annual with white-woolly herbage and leaves all basal. Leaves forming a basal rosette, pinnatifid with spinose teeth. Flowering scape 4 to 20 inches high; flowers in 1 to 4 interrupted whorls. Floral whorls subtended by leafy, spinescent-toothed bracts; densely white-woolly. Corolla pale lavender-blue; ¾ to 1 inch long; upper lip laciniate; lower lip fringed. Occurs in dry, sandy, gravelly places in the chaparral and also in the deserts. Blooms from March to June.

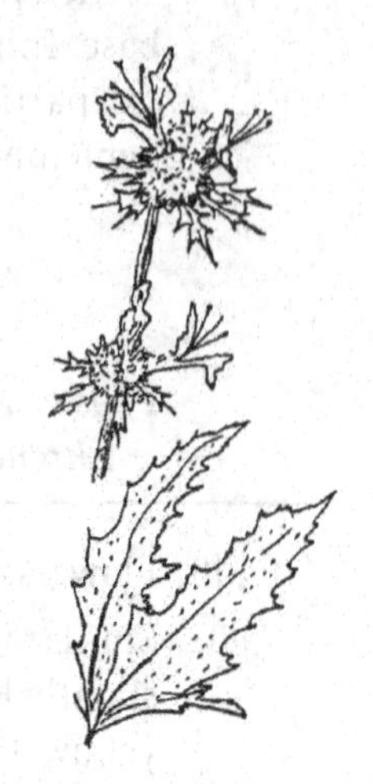

Lamiaceae

Salvia carduacea Benth. **Thistle Sage**

---

11. Plants finely pubescent. Leaves with rounded lobes. Corolla dark indigo blue

Erect annual with fine pubescence and basal leaves. Leaves forming a basal rosette, pinnatifid into rounded lobes. Flowers on a leafless stem, 4 to 20 inches tall in 1 to 3 compact, interrupted whorls. Whorls subtended by green to purplish bracts, broadly ovate and narrowing to a sharp awn tip. Calyx purple, about ⅜ inch long. Corolla dark purple-blue; ½ to ⅝ inch long. Common in dry, disturbed places throughout the chaparral. Blooms from March to June. Seeds of chia are edible and may be eaten raw or roasted and ground into a meal. They somewhat resemble poppy seeds. Native Americans dropped chia seeds into water for a flavorful drink.

Lamiaceae

Salvia columbariae Benth. **Chia**

---

10. Leaves opposite; toothed above. Flowers yellow, not in axillary whorls. Plant of wet areas
    See *Mimulus guttatus*, page 126.

9. Leaves alternate. Flowers in a dense terminal spike subtended by upper leaves (bracts) which are often colored

10. Floral bracts colored; red, pink to lavender

11. Bracts lavender tipped

Low, erect annual 4 to 16 inches high with villous or pubescent herbage. Leaves sessile, deeply pinnatifid into filiform, finger-like divisions. Flowers in a dense terminal spike; each flower subtended by 5 to 7 palmately lobed bracts ¼ to ¾ inch long with lavender to purple tips. Calyx 4-lobed; purple-tipped. Corolla barely visible, ½ to ¾ inch long, lavender with yellow and white markings. Owl's clover is common in early spring on grassy slopes and fields, sometimes barely visible above the young grass. Blooms March to May.

Orobanchaceae
Castilleja exserta (A. Heller) T.I. Chuang & Heckard

Purple Owl's Clover

---

11. Bracts red-tipped

12. Stem and leaves white, woolly

Perennial with grayish herbage, forming a bushy growth 1 to 2 feet high. Leaves are white-woolly, oblong to linear, and entire or 3-lobed above. Flowers are greenish but appear red because of scarlet tipped bracts and calyxes. Occurs on dry, open places and blooms from March to June.

Orobanchaceae
Castilleja foliolosa Hook. & Arn.

Woolly Paintbrush

---

12. Stem and leaves greenish, pubescent

Pubescent perennial with purplish stems that are more or less woody at the base; 12 to 20 inches high. Leaves lanceolate, palmately 3 to 5 lobed; frequently with axillary fascicles. Flowers in a terminal spike subtended by leaflike, fingered bracts which are scarlet tipped. Calyx 4-lobed; greenish; ¾ to 1 inch long; scarlet tipped. Corolla greenish, 1 to 1½ inches long; upper lip long with a thin, red margin. Fairly common in dry wooded areas of the chaparral. Blooms from March to May.

Orobanchaceae
Castilleja affinis Hook. & Arn.

Coast Indian Paintbrush

---

10. Floral bracts green

Pubescent annual 1 to 3½ feet high. Leaves alternate; pinnatifid into filiform lobes. Flowers white with purple lines; ½ to ⅝ inch long; upper lip arched, enclosing the anthers. Flowers in a 5 to 15 flowered head. Flowers subtended by green, hairy or bristly bracts ⅝ inch long with filiform lobes. Found on dry slopes. Blooms from May to August.

Orobanchaceae
Cordylanthus rigidus (Benth.) Jeps.
subsp. *setigerus* T.I. Chuang & Heckard

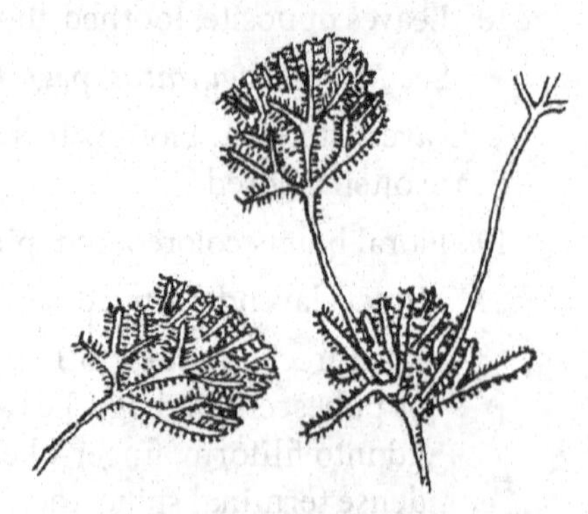

Bristly Bird's Beak

---

6. Leaves toothed or entire
7. Leaves opposite or basal (Leaves alternate, page 135)
8. Leaves mostly basal; heart-shaped. Flowers pansy-like

Slender-stemmed perennial 4 to 14 inches high. Leaves mostly basal, on petioles 1 to 2 inches long. Leaf blades heart-shaped to deltoid; crenate; ¼ to 1½ inches long. Flowers solitary, terminal on long peduncles which much exceed the leaves. Peduncles 5 to 6 inches long. Sepals 5; about ¼ inch long. Petals 5; irregular; golden yellow; ⅜ to ⅝ inch long. Lower petal slightly larger and somewhat sac-shaped or spurred. Upper petals red-brown on back; lower petals veined with brown. Common on grassy slopes and in fields in spring. Blooms from February to April.

Violaceae
Viola pedunculata Torr. & A. Gray

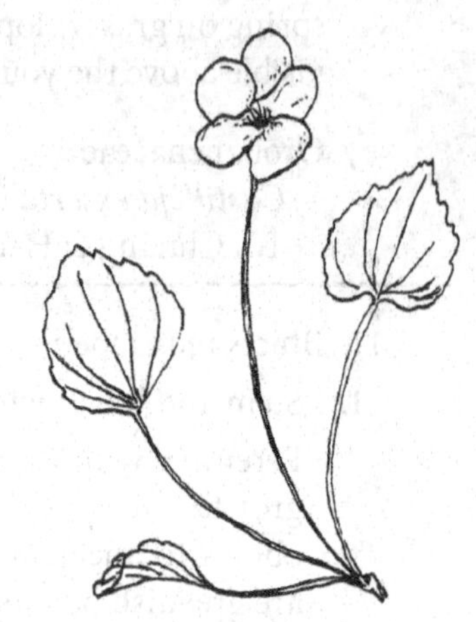

Johnny-Jump-Up

---

8. Leaves opposite
9. Flowers yellow
10. Plant not slimy
11. Pedicels longer than ½ the length of the calyx. Plants of moist or wet areas
12. Upper leaves connate; oval to rounded. Corolla ½ to 1½ inches long

Glandular pubescent perennial with watery stems 2 to 40 inches high. Leaves oval to rounded, coarsely dentate and more or less pinnatifid at the base; ½ to 3 inches long. Upper leaves sessile, connate. Flowering pedicels ¾ to over 2 inches long. Calyx angled, ½ to 1 inch long. Corolla bright yellow, ½ to 1½ inches long. Common plant in wet areas such as slow flowing streams. Blooms from March to August. Leaves may be eaten raw in salads.

Phrymaceae
Mimulus guttatus DC.

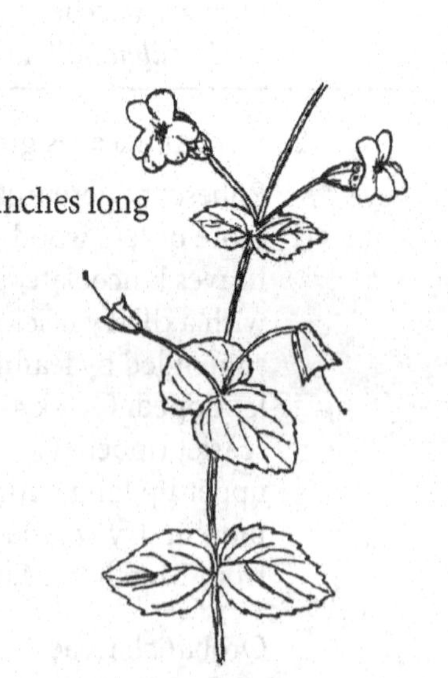

Seep-Spring Monkeyflower

---

12. Leaves lanceolate to oblong, not connate. Corolla with maroon spots at base, ¼ inch long

Low, villous annual 2 to 14 inches high. Leaves lanceolate to oblong, sessile, entire, ⅜ to over 1 inch long. Flowering pedicels ⅜ to ⅝ inch long. Calyx about ¼ inch long. Corolla yellow, almost regular, a little over ¼ inch long with maroon spots at the base. Occasional plant of moist, gravelly places such as partially dried stream beds. Blooms from April to September.

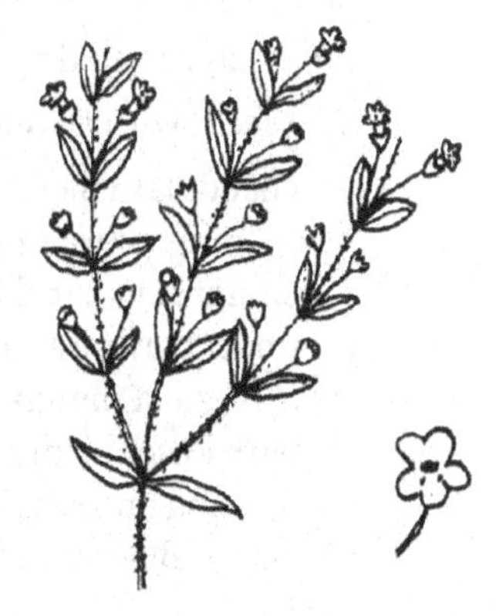

Phrymaceae
Mimulus pilosus (Benth.) S. Watson

Downy Monkeyflower

---

11. Pedicels short; less than ½ the length of the calyx. Plant of dry areas

Sticky pubescent annual with unbranched stems 4 to 32 inches high. Basal leaves ovate; upper leaves lanceolate, ¾ to 2½ inches long; sessile, entire. Flowering pedicels ¼ inch or less. Calyx ½ to ¾ inch long. Corolla lemon-yellow, two-lipped; 1 to 2 inches long. Inhabitant of dry, rocky, exposed places. Blooms in spring from April to June.

Phrymaceae
Mimulus brevipes Benth.

Wide-Throated Yellow Monkeyflower

---

10. Plant slimy

Extremely slimy, villous perennial with creeping stems 4 to 12 inches long. Leaves ovate, toothed; ¼ to 2 inches long. Pedicels ⅜ to 2 inches long, slender. Calyx slightly less than ½ inch long. Corolla ¾ to 1 inch long, with a long cylindrical tube. Inhabitant of wet places. More common in the higher mountains than in chaparral. Occasionally found around Lake Sherwood.

Phrymaceae
Mimulus moschatus Lindl.

Musk Monkeyflower

---

9. Flowers not yellow

10. Leaves entire or minutely toothed.

(Leaves distinctly toothed, page 131)

11. Herbage with distinct mint odor

12. Weed, with strong vinegar odor. Flowers in axils of upper leaves

Glandular villous annual 4 to 24 inches high, with a strong vinegar odor. Leaves opposite, lanceolate, entire; ¾ to 2 inches long with prominent veins on the under surface and acute to acuminate apex. Flowers in clusters in the axils of the upper leaves. Corolla light blue, lower lip drawn back and stamens long exserted, forming an arch. Common weed in open or disturbed places. Blooms in the summer from August to October.

Lamiaceae
Trichostema lanceolatum Benth.

Vinegar Weed

---

12. Herb, with pleasant mint odor. Flowers in a terminal head subtended by bracts

13. Bracts and under surface of leaves white-woolly

Perennial, 8 to 20 inches high with pubescent stems and opposite leaves; slightly woody at the base. Leaves lanceolate entire, glabrous above and white-woolly beneath; ¾ to 1½ inches long on petioles up to ⅜ inch long. Flowers almost regular, white to lavender; ⅝ inch long. Stamens 4; exserted. Flowers in heads 1 to 1½ inches across subtended by woolly bracts ⅜ inch long. Occurs on dry slopes from San Diego to Santa Barbara. Blooms from July to September. **Felt-Leaved Monardella**, *Monardella hypoleuca* A. Gray subsp. *lanata* (Abrams) Munz, occurs in east San Diego County, has villous branches and leaves which are pubescent to slightly woolly above. Blooms from June to July.

Lamiaceae
Monardella hypoleuca A. Gray

White-Leaf Monardella

---

13. Bracts green, with purplish tips. Under surface of leaves sparsely pubescent

Erect annual with a pleasant mint odor. Stem glabrous below and finely pubescent above; 8 to 20 inches high. Leaves opposite: lanceolate entire; ½ to 2 inches long on petioles up to ½ inch long. Flowers in dense heads ½ to 1 inch across, subtended by membranous bracts, lanceolate ovate with acute purplish tips. Flowers rose to purple; about ½ inch long. Stamens 4; exserted. Occurs in dry places. Blooms from May to August.

Lamiaceae
Monardella breweri A. Gray
subsp. *lanceolata* (A. Gray) A.C. Sanders & Elvin

Mustang Mint

---

11. Herbage without distinct mint odor

12. Flowers red; long tubular

Glabrous, glaucous perennial somewhat woody; 1 to 4 feet high. Leaves opposite, entire, lanceolate below; auriculate clasping above. Flowers in a compact panicle. Corolla tubular, 1 to 1¼ inches long; scarlet, almost regular. Common in dry sandy soil. Blooms from April to July.

Plantaginaceae
Penstemon centranthifolius Benth.

Scarlet Bugler

---

12. Flowers white to blue-violet; not long tubular

13. Flowers in interrupted whorls

Annual, slightly pubescent; 8 to 20 inches tall with greenish purple stems. Leaves opposite, glabrous, lanceolate, slightly toothed; ⅜ to 3 inches long. Flowers in axillary whorls subtended by leaflike bracts. Calyx 5-lobed; villous, teeth acute to acuminate. Corolla ⅝ to ¾ inch long; upper lip lilac; lower lip blue-violet. Inhabitant of dry, shaded areas of the chaparral. Some years it may be quite abundant. Blooms from March to July.

Plantaginaceae
Collinsia heterophylla Graham

Chinese-Houses

---

13. Flowers axillary, or in lateral or terminal racemes

14. Flowers small, ⅛ to ¾ inch long. Leaves ovate to lanceolate

15. Flowers ⅛ to ⅜ inch long, in lateral racemes arising from leaf axils

Somewhat succulent herb with stems 4 to 40 inches high. Leaves opposite, sessile, oblong to lanceolate, minutely toothed. Flowers in several lateral racemes arising from leaf axils. Racemes many-flowered. Corolla 4-lobed, slightly irregular, pale blue-violet, ⅛ to ¼ inch across. Stamens 2. Plant of wet places, such as along streams. Blooms from May to September. Also growing in wet places along streams is **American Brooklime,** *Veronica americana* Benth. It is similar except that the racemes are fewer flowered, and the corolla is larger; ¼ to ⅜ inch across. Leaves and stem of both species are edible and may be eaten raw or cooked.

Plantaginaceae
Veronica anagallis-aquatica L.

Water Speedwell

---

15. Flowers ⅜ to ¾ inch long, axillary or in terminal racemes
  16. Flowers white or lavender to rose-red
    17. Flowers white to pale lavender, ⅜ to ½ inch long

Erect annual to 1 to 4 feet tall, glabrous below the inflorescence. Leaves ovate to lanceolate; the lower ones opposite; the upper, alternate; ¾ to 3½ inches long. Upper stem bearing slender, tortuous branchlets. Flowers in a dense, terminal raceme. Calyx villous. Corolla white to pale lavender; 2-lipped, with a sac-shaped base; the upper lip closing the throat of the flower. Occasional in dry, disturbed places or common following a burn. Blooms from April to July.

Plantaginaceae
Antirrhinum coulterianum A. DC.

White Snapdragon

---

17. Flowers pink to rose-red *Antirrhinum multiflorum*, page 136.
  16. Flowers blue or blue-violet
    17. Stems twining. Pedicels 1½ to 3 inches long

Annual with twining stems, and long, capillary pedicels. Stems 1 to 3½ feet long. Leaves ovate, ¼ to ¾ inch long; lower ones petioled; and opposite below; alternate above. Pedicels twining, 1½ to 3 inches long. Flowers blue, ½ inch long solitary in the axils of leafy bracts. Fruit a glabrous, globose capsule, ¼ inch in diameter. Occurs on dry slopes, particularly after a burn. Blooms from March to May.

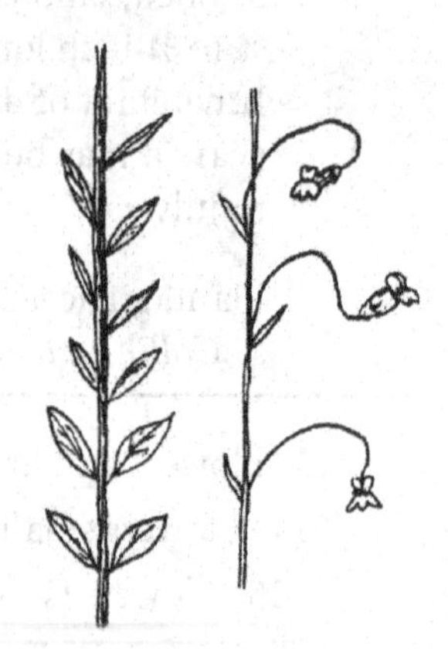

Plantaginaceae
Antirrhinum kelloggii Greene

Twining Snapdragon

---

17. Stems not twining. Pedicels ¼ to ¾ inch long

Erect annual or biennial; branched below. Stems supported by curved branchlets or at least possessing them. Stems glandular pubescent; 4 to 40 inches long. Leaves alternate; ovate; entire; ¼ to 1½ inches long. Flowers blue-violet, ⅜ to ½ inch long; in leafy-bracted racemes. Fruit a glandular, pubescent capsule; ¼ inch in diameter. Found in dry, disturbed places from Santa Barbara south. Blooms from March to July.

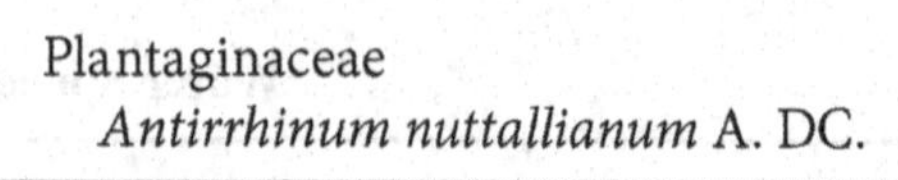

Plantaginaceae
Antirrhinum nuttallianum A. DC.

Violet Snapdragon

---

14. Flowers larger, 1 to 1½ inches long. Leaves linear; fascicled

Perennial, slightly woody at the base, with stems 12 to 20 inches high. Leaves linear, glabrous, much fascicled; ¾ to 2 inches long. Flowers in a terminal raceme; flowers subtended by reduced leaves. Corolla showy, 2-lipped, 1 to 1¼ inches long, blue-violet, on pedicels averaging ½ inch long. Common on dry hillsides. Blooms from April to July.

Plantaginaceae
Penstemon heterophyllus Lindl.

Foothill Penstemon

---

10. Leaves distinctly toothed

11. Flowers in axillary whorls or in a spike of interrupted whorls (Flowers not in whorls, page 133)

12. Leaves roundish, wrinkled. Calyx with 10 spiny teeth. Flowers white

Perennial herb 8 to 24 inches high with square stem, but no mint odor. Herbage densely white-woolly, particularly when young. Leaves roundish, crenate, wrinkled or rugose, tomentous beneath. Calyx sharply 5 to 10 toothed, becoming a prickly bur when dry. Corolla small, white, ¼ inch long; in dense, axillary whorls. Common weed in waste laces and old fields. Blooms during spring and summer. Tea made from dried leaves of this plant is said to be useful as a laxative or tonic. This is also the source of horehound candy, which is popular for coughs and sore throat.

Lamiaceae
Marrubium vulgare L.

Horehound

---

12. Leaves ovate to lanceolate. Calyx 5-toothed or lobed

13. Flowers dark red, 1 to 1½ inches long

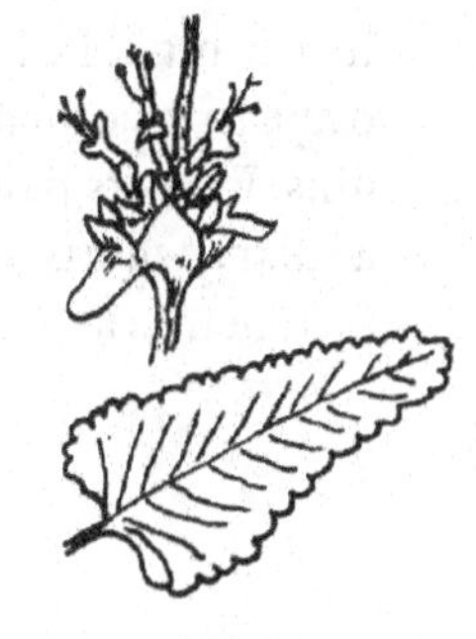

Coarse perennial with stout stem and villous, glandular herbage; 1 to 2½ feet high. Leaves oblong to sagittate, green rugose above; gray-white tomentose beneath. Margin irregularly crenate. Petioles long villous. Blades 3 to 8 inches long. Flowers in compact interrupted whorls subtended by ovate to lanceolate bracts which are purplish. Corolla dark red; 2-lipped; 1 to 1½ inches long. Stamens long exserted. Sometimes abundant on shaded, grassy slopes, particularly after a wet winter. Blooms from April to July.

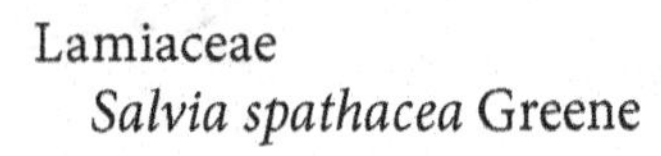

Lamiaceae
Salvia spathacea Greene

California Hummingbird Sage

---

13. Flowers not dark red. Corolla less than 1 inch long

14. Flowers small, less than ½ inch long, almost regular. Leaves ovate-lanceolate with rounded to acute base. Aromatic herb with a pleasant odor

15. Leaves ⅜ to ¾ inch long, covered with grayish-white hairs

Aromatic, perennial herb with pubescent stems 8 to 20 inches high. Leaves ⅜ to ¾ inch long, ovate to elliptic with toothed margins and covered with short hairs. Flowers lavender, in interrupted whorls. Occurs in low, moist areas. Blooms from June to September. The oil is **TOXIC** if ingested.

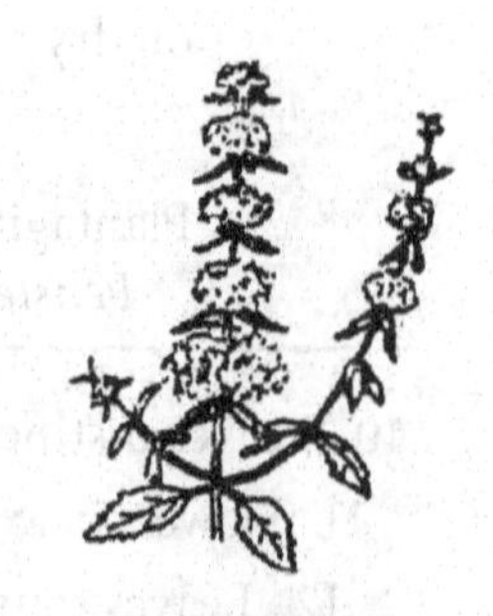

Lamiaceae
Mentha pulegium L.

Pennyroyal

---

15. Leaves ¾ to 3 inches long; not covered with grayish-white hairs

16. Stem glabrous. Flowers in a terminal spike of interrupted whorls

Aromatic, perennial herb with glabrous stems, 1 to 4 feet high. Leaves opposite; ovate to lanceolate; ¾ to 2¼ inches long; glabrous to pubescent. Flowers pale lavender in terminal spikes of interrupted whorls. Moist fields; from July to October. Leaves may be used for tea or for flavoring.

Lamiaceae
Mentha spicata L.

Spearmint

---

16. Stem pubescent. Flowers in axillary whorls

Pubescent perennial; aromatic; with stems 4 to 32 inches high. Leaves opposite; ¾ to 3 inches long; ovate to lanceolate; light green, with toothed margins. Flowers pale pink, lilac to purple, occurring in axillary whorls. Grows in moist places, from July to October. Leaves may be used for tea or for flavoring.

Lamiaceae
Mentha arvensis L.

Field Mint

---

14. Flowers larger, distinctly 2-lipped. Corolla ½ to ¾ inch long. Leaves ovate with cordate base

15. Plant rough hairy. Corolla purple with white markings. Plant of dry slopes

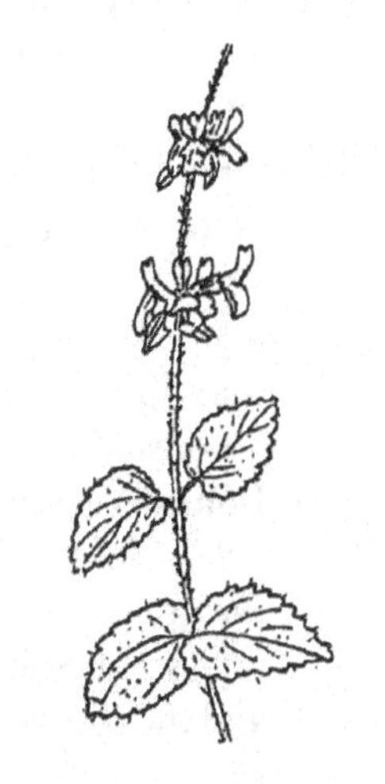

Slender, glandular pubescent, weak-stemmed herb 16 to 32 inches high. Leaves ovate, 1 to 7 inches long, broadly crenate with truncate or slightly heart-shaped base. Flowers in whorls of 6 flowers. Calyx about ¼ inch long, spinose toothed. Corolla lavender to red-violet with white markings or mottling on the lower lip. Corolla ½ to ¾ inch long. Occasional on dry slopes and in canyons. Blooms from April to September.

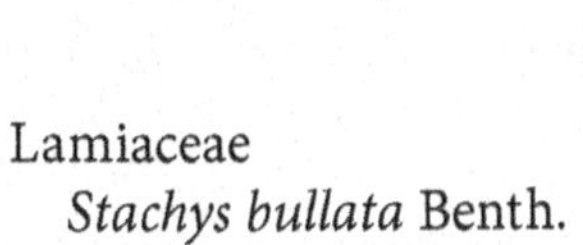

Lamiaceae
Stachys bullata Benth.

California Hedge Nettle

---

15. Plant white-woolly. Corolla white with purple veins. Plant of moist places

Stout, branched, white-woolly herb to 3½ feet tall. Leaves ovate, crenate, 1 to 5 inches long; tomentose beneath. Flowers in long spikes 4 to 8 inches long. Calyx woolly, about ½ inch long. Corolla white to pink with purple veins; ½ to ⅝ inch long. Inhabitant of moist places, sometimes quite abundant. Blooms in the summer from May to October.

Lamiaceae
Stachys albens A. Gray

White Hedge Nettle

---

11. Flowers not in axillary whorls or in a spike of interrupted whorls

12. Flowers small, less than ½ inch long

13. Low, trailing herb. Flowers white

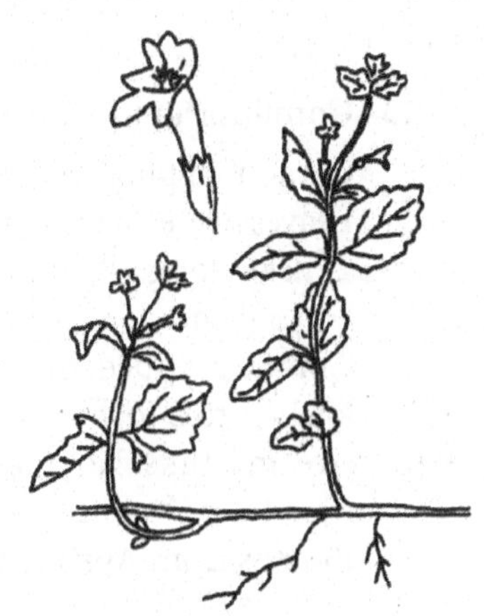

Trailing evergreen herb with stems 8 to 24 inches long. Leaves round-ovate, short petioled; ½ to 1 inch long; pubescent with crenate margins. Flowers white to more or less purple; ¼ inch long on pedicels ⅜ to ⅝ inch long. Calyx tubular. Flowers solitary in axils of leaves. In shaded woods from Los Angeles County northward. Dried leaves of yerba buena may be used to make a tea which is supposed to stimulate digestion.

Lamiaceae
Clinopodium douglasii (Benth.) Kuntze

Yerba Buena

---

13. Erect plant, 1 to 6 feet high

14. Flowers purple, in terminal spikes. Corolla small, less than ¼ inch across, almost regular. Plant 1 to 3 feet high

Villous to hirsute perennial, much branched, 1 to 3 feet high. Leaves opposite, ovate, coarsely toothed to lobed, ¾ to 2¼ inches long. Flowers in narrow, terminal spikes, 2 to 8 inches long. Calyx 5-toothed. Stamens 4. Corolla 5-lobed, slightly irregular; dark blue to blue-purple; less than 3/16 inch across. Common weed in moist places. Blooms May to September. Also in moist places may be **Robust Vervain**, *Verbena lasiostachys* Link var. *scabrida* Moldenke, which is similar to the above except that leaves are greener and scabrous, rather than gray, pubescent. Spikes are dense, 1¼ to 4 inches long in fruit, and the corolla is smaller; ⅛ inch across.

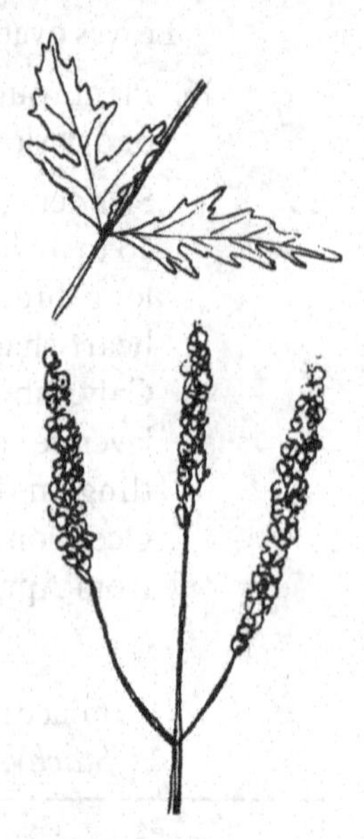

Verbenaceae
Verbena lasiostachys Link
var. *lasiostachys*

Western Vervain

---

14. Flowers red-brown, in loose clusters. Corolla ¼ to ⅜ inch long, distinctly irregular. Plant 3 to 6 feet tall

Tall, coarse herb 3 to 6 feet tall. Stems somewhat pubescent. Leaves opposite, sharply toothed, ovate or triangular ovate, 1 to 3 inches long. Flowers in loose clusters arising from leaf axils. Corolla maroon or red-brown; ¼ to ⅜ inch long. Fairly common in more or less damp areas. Blooms from February to July.

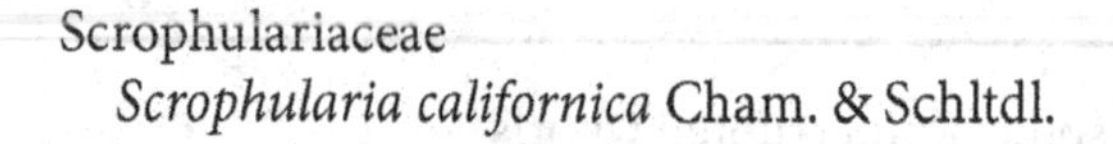

Scrophulariaceae
Scrophularia californica Cham. & Schltdl.

California Figwort

---

12. Flowers larger, showy, ½ to 2 inches long

13. Corolla bright red. Plant of wet areas, such as stream beds

Glandular, pubescent perennial 8 to 32 inches high. Leaves ovate, sharply toothed, opposite, sessile; ⅜ to 2 inches long. Flowers solitary in leaf axils on pedicels 2 to 3 inches long. Calyx tubular, ¾ to 1¼ inches long. Corolla 2-lipped, orange-red or scarlet with some yellow; 1½ to 2 inches long. Occurs in moist areas such as stream banks. Not too common in the Santa Monica Mountains but may be abundant during wet years. Blooms from April to October.

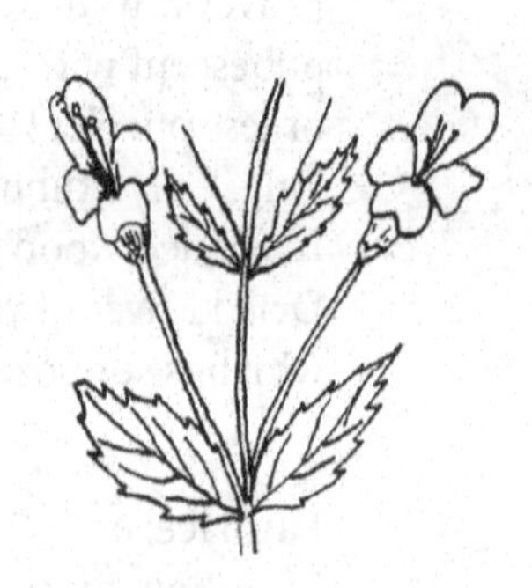

Phrymaceae
Mimulus cardinalis Benth.

Scarlet Monkeyflower

---

13. Corolla blue to blue-violet or red-violet

14. Plant low, 1 to 8 inches high. Leaves ⅜ to ⅝ inch long

Erect, perennial herb with glabrous to villous stems; 1 to 8 inches high. Leaves opposite; ovate; with crenate margins and ⅜ to ⅝ inch long. Flowers dark purple, ½ to ¾ inch long; 1 to 3 flowers on short pedicels in axils of leaves. Grows on slopes in southern California. Blooms from April to May.

Lamiaceae
Scutellaria tuberosa Benth.

Skullcap

---

14. Plant taller, 2 to 4 feet high. Leaves ¾ to 4 inches long. Upper leaves connate

Erect, glabrous herb 32 to 48 inches high. Leaves opposite, sessile, coarsely toothed, ¾ to 4 inches long; upper leaves connate perfoliate. Flowers showy, in loose panicles arising from leaf axils. Calyx about ¼ inch long. Corolla red-violet to purple; 1 to 1½ inches long. Base of corolla narrow, then becoming abruptly expanded. Common in dry disturbed places. Blooms April to June.

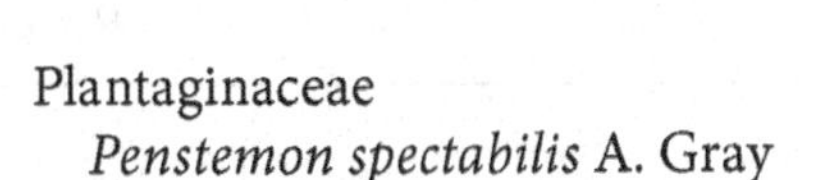

Plantaginaceae
Penstemon spectabilis A. Gray

Showy Penstemon

---

7. Leaves alternate

8. Flowers in a dense, terminal spike subtended by colored or green leaflike bracts

9. Bracts red-tipped

Pubescent annual; 1 to 5 feet high. Leaves ¾ to 3 inches long; linear-lanceolate with entire margins. Flowers dull red-yellow, 1 to 1¼ inches long. Sepals and bracts entire. Fruit a capsule ½ inch in diameter. Frequent in wet places. Blooms from May to September.

Orobanchaceae
Castilleja minor (A. Gray) A. Gray
subsp. *spiralis* (Jeps.) T.I. Chuang & Heckard

Lesser Paintbrush

---

9. Bracts green

Branched, decumbent annual with pubescent and more or less gland-tipped stems 8 to 16 inches long. Leaves alternate; oblong to lanceolate; ¼ to 1 inch long. Flowers ⅝ to ¾ inch long; purple on the lower lip. Fruit a capsule ¼ inch long. Found on coastal salt marshes. Blooms from May to October.

Orobanchaceae
Chloropyron maritimum (Benth.) A. Heller subsp. *maritimum*

Salt Marsh Bird's-Beak

---

8. Flowers in an elongate raceme
9. Flowers white, lavender, pink to rose-colored
10. Flowers white to pale lavender
See *Antirrhinum coulterianum*, page 130.
10. Flowers pink to rose-colored

Glandular pubescent annual, 2 to 5 feet high. Upper leaves alternate, lanceolate. Flowers in a terminal raceme. Corolla pink to rose-colored, 2-lipped, ½ to ¾ inch long. Grows on dry slopes of the chaparral, particularly following a fire.

Plantaginaceae
Antirrhinum multiflorum Pennell

Rose Snapdragon

---

9. Flowers blue or blue violet
10. Stems twining. Pedicels 1½ to 3 inches long
See *Antirrhinum kelloggii*, page 130.
10. Stems not twining. Pedicels ¼ to ¾ inch long
See *Antirrhinum nuttallianum*, page 130.

SECTION III
COMPOSITES—SUNFLOWER FAMILY (ASTERACEAE)

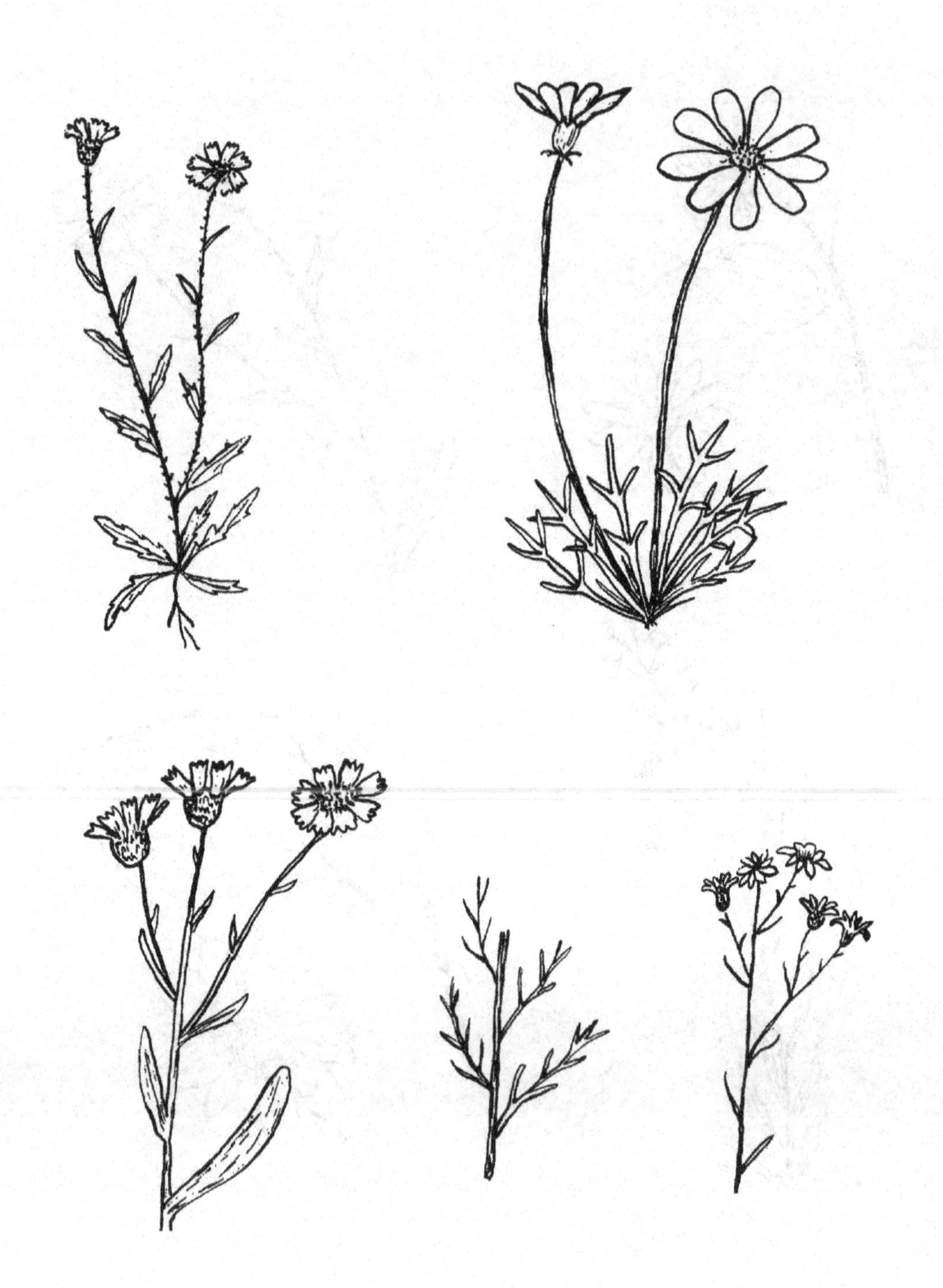

1. Flowering heads with ray flowers (No ray flowers, see page 163)
   2. Heads with both ray and disk flowers (Only ray flowers, page 156)
      3. Shrubs, woody at the base. Ray flowers yellow (Herbs, see page 143)
         4. Leaves pinnatifid; at least below
            5. Flowers large, showy, 1½ to 3 inches across. Lower stem with treelike trunk

Stout, glabrous shrub 1½ to 10 feet tall with tree-like trunk below and up to 5 inches thick. Upper portion branched; leaves at ends of branches. Leaves pinnatifid into filiform segments. Heads large and showy, 1½ to 3 inches across; terminal on scape-like peduncles. Ray flowers 10 to 16, bright yellow. Disk flowers yellow; numerous. Occurs on rocky cliffs and bluffs along the coast. Blooms from March to May.

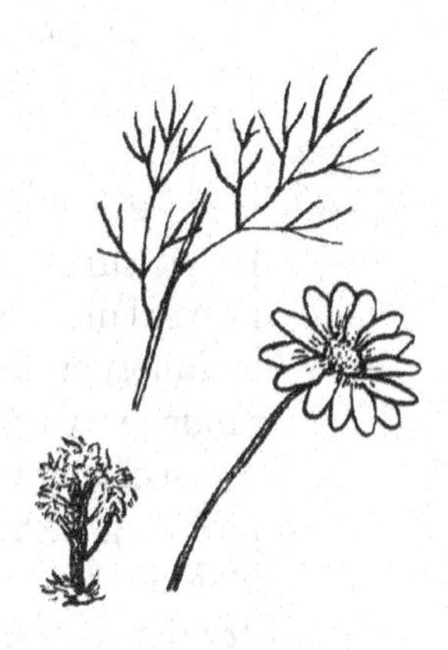

Asteraceae
Leptosyne gigantea Kellogg

Giant Tickseed

---

5. Flowers smaller, less than 1½ inches across
   6. Herbage more or less woolly. Leaves pinnatifid into linear segments
      7. Plant 2 to 5 feet tall. Ray flowers ½ inch long

Branched, bushy shrub 3 to 5 feet high with tomentose herbage, particularly when young. Leaves gray-green tomentose; pinnatifid into linear segments and often fascicled. Heads showy and numerous in open clusters at tips of branches. Rays yellow, about ½ inch long. Common plant in gravelly washes and dry stream beds. Mostly away from the coast. Blooms June to October.

Asteraceae
Senecio flaccidus Less.
var. *douglasii* (DC.) B.L. Turner & T.M. Barkley

Threadleaf Ragwort

---

7. Plant 1/2 to 2 feet tall. Rays 3/16 inch long

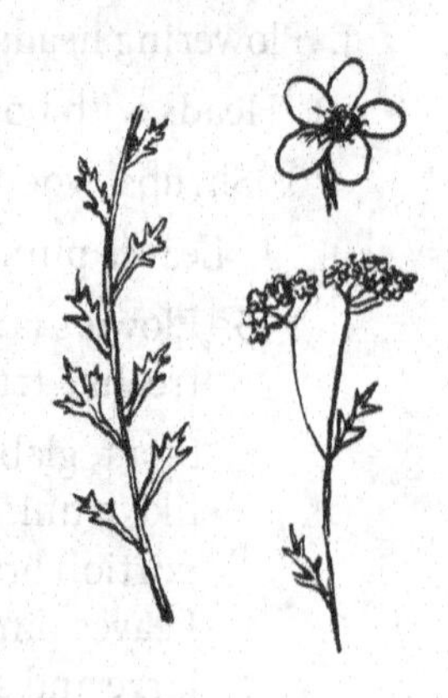

Shrub 8 to 24 inches high with numerous erect branches from a woody base. Herbage white tomentose, particularly when young. Leaves pinnate into 3 to 5 linear lobes; green above, tomentose beneath; margins revolute. Heads small and numerous, in dense terminal clusters. Rays 4 to 6; orange-yellow; slightly more than 1/8 inch long. Very common plant on brushy chaparral slopes. Blooms April to August.

Asteraceae

Eriophyllum confertiflorum (DC.) A. Gray **Golden-Yarrow**

---

6. Herbage sandpapery; not woolly. Upper leaves lanceolate, deeply toothed

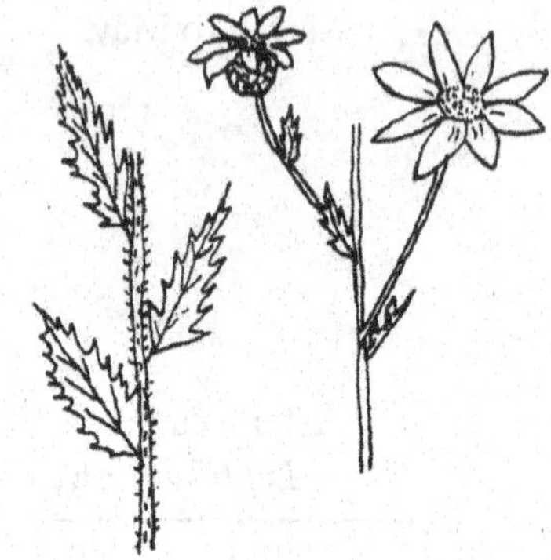

Hispid shrub 2 to 4 feet high with alternate leaves. Leaves lanceolate, 3/4 to 3 inches long; deeply toothed to almost entire. Leaves rather veiny beneath, with sandpapery texture on short petioles. Heads in a terminal cluster with 8 to 13 yellow ray flowers, about 1/2 inch long. Achenes with appressed hairs and a pappus of deciduous, awn-tipped scales and awns. Occurs on dry slopes in southwestern San Diego County. Blooms from February to June.

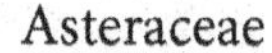

Asteraceae

Bahiopsis laciniata (A. Gray) E.E. Schill. & Panero **San Diego Viguiera**

---

4. Leaves not pinnatifid
5. Leaves linear to filiform; often fascicled. Herbage resinous
6. Ray flowers 8 to 30
7. Flowers in spring
8. Leaves linear; 1/16 to 1/8 inch wide

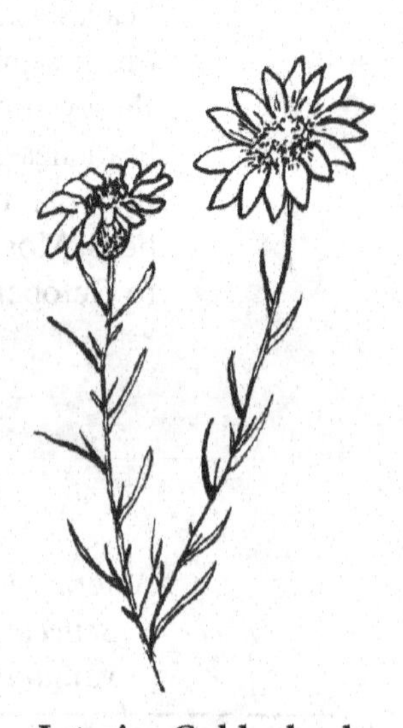

Resinous shrub 1½ to 5 feet high. Leaves alternate; crowded and fascicled; linear, 3/8 to 1½ inches long and 1/8 inch wide or less. Stems glandular and finely pubescent below the heads. Heads numerous, 1/4 to 1/2 inch high in terminal clusters. Rays 13 to 18; yellow; 3/8 to 5/8 inch long. Involucral bracts in 2 to 3 series with ciliate margins. Achenes silky, hairy, with a pappus of many white capillary bristles which are soon deciduous. Found in rocky or sandy soil on arid slopes and banks. Blooms March to May.

Asteraceae

Ericameria linearifolia (DC.) Urbatsch & Wussow **Interior Goldenbush**

---

8. Leaves filiform; pine-like; less than 1/16 inch wide

Stout shrub 2 to 8 feet high with a trunk-like main stem. Leaves filiform, ⅝ to 1¼ inches long, resin dotted and with axillary fascicles. Flowering heads in both spring and fall. Spring heads solitary and few in number with 15 to 30 yellow ray flowers. Fall heads smaller, in leafy bracted panicles, with 5 to 10 ray flowers. Disk flowers about 12 to 18. Pine-bush grows in dry sandy areas from northern Los Angeles County to San Diego. Blooms from April to July and again from September to January.

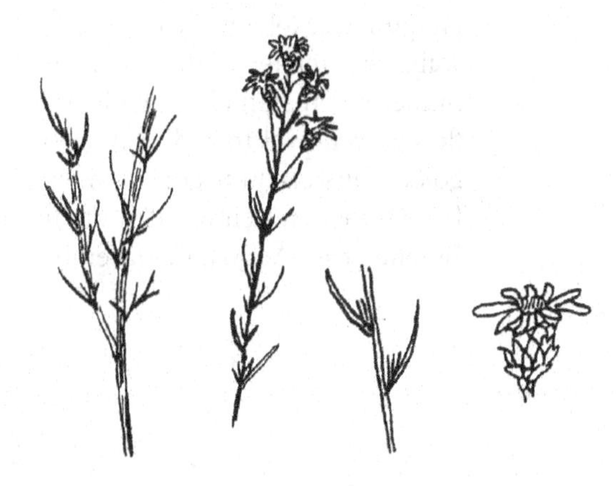

Asteraceae
Ericameria pinifolia (A. Gray) H.M. Hall

Pine-Bush

---

7. Flowers in autumn. Rare plant in Santa Susana Mountains

Resinous and fragrant shrub 2 to 3½ feet high. Leaves small and crowded, linear, entire; sticky resinous. Heads somewhat showy, solitary on long peduncles. Involucres ¼ inch high. Ray flowers 8; yellow; ¼ inch long. Disk flowers 18 to 23. Rare plant in the Santa Susana Mountains, usually found among rocks. Blooms in the late summer and fall; July to October. **RARE.**

Asteraceae
Deinandra minthornii (Jeps.) B.G. Baldwin

Santa Susana Tarplant

---

6. Ray flowers 1 to 6
7. Leaves short; ⅛ to ½ inch long
8. Plant of sand dunes near or on the coast; Los Angeles County and north. Achenes glabrous

Compact, dense, much branched shrub 1 to 3½ feet tall, with resinous herbage. Leaves linear, densely fascicled, ⅛ to ⅜ inch long. Heads numerous at tips of short branches. Involucral bracts green tipped, distinctly imbricated. Ray flowers 2 to 6; yellow. Disk flowers 8 to 14. Common shrub on dunes and sandy areas near the coast. Blooms from August to November.

Asteraceae
Ericameria ericoides (Less.) Jeps.

Mock Heather

---

8. Plant of dry soil; inland; Ventura County to the desert border. Achenes pubescent

Resinous shrub 1½ to 4½ feet high with glandular, punctate foliage. Leaves fascicled filiform and somewhat terete; ⁵⁄₁₆ to ½ inch long. Flowering heads numerous, in panicles. Involucres ¼ inch high. Ray flowers yellow, 1 to 6; disk flowers 5 to 10. Palmer's golden-bush is quite common on dry, flat areas from southern Ventura County to the desert border. Blooms from August to December.

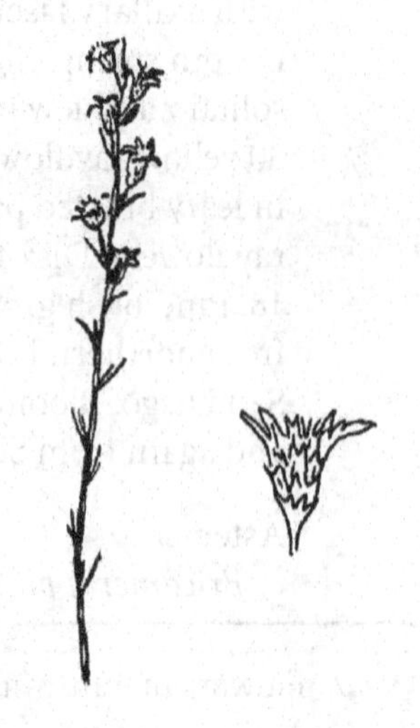

Asteraceae
Ericameria palmeri (A. Gray) H.M. Hall
var. *pachylepis* (H.M. Hall) G.L. Nesom **Thickbracted Goldenbush**

---

7. Leaves longer; ½ to 1¼ inches long
See *Ericameria pinifolia*, page 141.

5. Leaves wider; not linear or filiform

6. Leaves deeply, irregularly toothed. Herbage scabrous
See *Bahiopsis*, page 140.

6. Leaves shallowly toothed to entire

7. Leaves lanceolate. Disk flowers red-brown

Glabrous or sparingly pubescent perennial, woody at base and growing in bushy clumps 2 to 5 feet high. Leaves ovate to lanceolate, acute, 1 to 2½ inches long; ⅜ to 1¼ inches wide; alternate, entire, petiolate. Heads showy, 2 or more inches across. Rays yellow; disk red-brown. Involucral bracts densely yellow or tawny villous; imbricated; ovate, with acute apex. On chaparral slopes and coastal bluffs. Blooms from February to June.

Asteraceae
Encelia californica Nutt. **California Brittlebush or Bush Sunflower**

---

7. Leaves deltoid to ovate. Disk flowers yellow

Tall, branched, glabrous herb, slightly woody at the base, entire or crenate dentate. Leaves thin, 2 to 6 inches long. Heads solitary on slender peduncles, ¾ to 2½ inches long. Outer involucral bracts leaf-like and spreading. Inner bracts circular with a rounded apex and thin or scarious. Rays 13 to 21, bright yellow, ½ to ¾ inch long. Heads mostly 1½ to 2 inches across. Common plant on shaded slopes and along moist stream banks. Blooms from February to September.

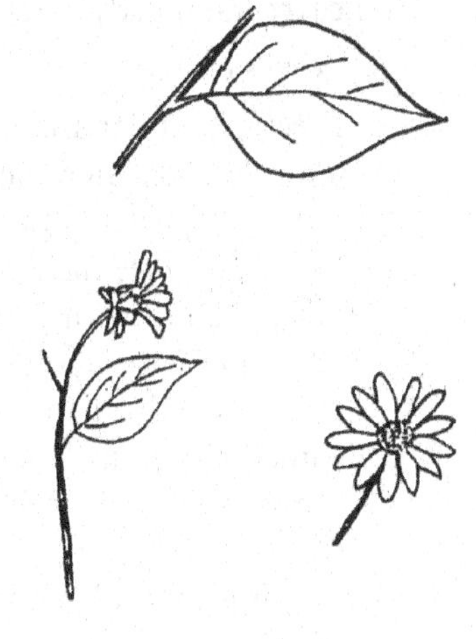

Asteraceae
Venegasia carpesioides DC.

Canyon Sunflower

---

3. Herbs; not woody at the base. Ray flowers yellow, purple, or white
4. Ray flowers yellow or yellow/white (Ray flowers not yellow, page 153)
5. Ray flowers yellow with white tips

Short annual 4 to 12 inches high, with glandular pubescent to short, hispid herbage. Leaves alternate; basal ones pinnatifid to dentate; upper ones entire. Heads terminal, showy, 1 to 2 inches across. Rays ¼ to ½ or more inches long; yellow with a white tip. Disk flowers numerous; 30 to 100 flowers. Common in grassy fields. Blooms from April to June.

Asteraceae
Layia platyglossa (Fisch. & C.A. Mey.) A. Gray

Tidy-Tips

---

5. Ray flowers without white tips
6. Leaves pinnatifid or pinnately parted (Leaves toothed or entire, p. 145)
7. Leaves mostly basal

Erect, glabrous annual 4 to 24 inches high. Leaves mostly basal, pinnatifid into linear lobes. Heads solitary; terminal on a scape; showy, with golden yellow ray and disk flowers. Heads ¾ to 2 inches across. Outer involucral bracts linear; inner ones ovate. Common on dry, gravelly hillsides. More abundant in the northern deserts, sometimes producing spectacular coloring. Blooms from March to May.

Asteraceae
Leptosyne bigelovii (A. Gray) A. Gray

Bigelow's Tickseed

---

7. Leaves not mostly basal

8. Leaves opposite

Erect, glandular and more or less villous annual 4 to 12 inches high. Leaves opposite; 1 or 2 times pinnately parted into narrow or filiform lobes. Leaves ⅜ to 1½ inches long, without petioles. Heads yellow with 8 to 15 ray flowers, $\frac{3}{16}$ to ⅜ inch long. Involucral bracts ovate with a midrib. Achenes with a pappus of 8 to 2 small scales or sometimes none. In sunny, sandy spots from March to May. *Lasthenia coronaria* is not nearly as common as **California Goldfields**, *Lasthenia californica* Lindl. subsp. *californica*.

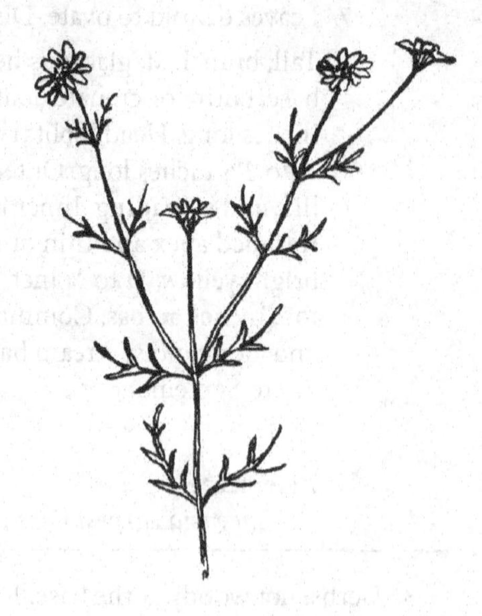

Asteraceae
Lasthenia coronaria (Nutt.) Ornduff **Royal Goldfields**

---

8. Leaves alternate

9. Leaves two times pinnatifid. Escape from cultivation

Erect, glabrous annual 8 to 40 inches high. Leaves alternate, 2 times pinnatifid into oblong divisions. Heads showy, ¾ to 1½ inches across. Ray flowers either all yellow or mostly white with yellow base each about ¾ inch long with a toothed apex. Garland daisy blooms from April to August. Naturalized in fields and waste places.

Asteraceae
Glebionis coronaria (L.) Spach **Garland Daisy**

---

9. Leaves once pinnatifid

10. Outer flowers, enlarged disk flowers; deeply lobed and ray-like

Erect annual 4 to 24 inches high; slightly woolly when young, but soon becoming glabrous. Leaves pinnatifid into narrow lobes. Heads showy, ½ to 1½ inches across. Flowers all disk the outer ones much enlarged with a spreading palmately lobed limb which surpasses the disk. Inhabitant of sandy areas, particularly along the coast. Blooms from March to May.

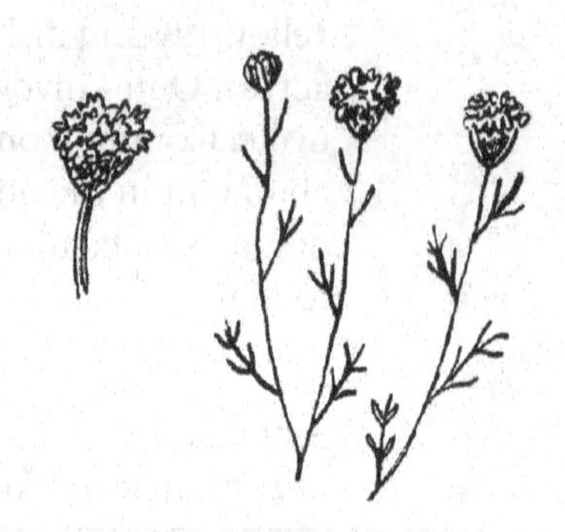

Asteraceae
Chaenactis glabriuscula DC. **Yellow Pincushion**

---

10. Outer flowers true ray flowers, not deeply lobed

11. Annual 4 to 16 inches high. Rays less than ⅜ inch long

Glabrous annual, 4 to 20 inches high. Leaves linear to lanceolate; pinnatifid, or upper leaves toothed; ⅜ to 2¼ inches long with an auriculate base. Rays yellow, ⅜ inch long. Involucral bracts black-tipped. Found in dry, open places. Blooms March to May.

Asteraceae
Senecio californicus DC. **California Ragwort**

---

11. Perennial 16 to 40 inches high. Rays ⅜ to ¾ inch long

Perennial herb 1 to 3 feet high; glabrous or with tufts of wool in leaf axils. Leaves alternate, pinnatifid into 1 to 9 pairs of segments. Terminal segment, larger and toothed. Upper leaves reduced. Heads with yellow rays ⅜ to ¾ inch long. Involucral bracts in one series. Achenes with a pappus of white bristles. Found on brushy slopes from Los Angeles County north. Blooms from April to June.

Asteraceae
Packera breweri (Burtt Davy) W.A. Weber & Á. Löve **Brewer's Ragwort**

---

6. Leaves toothed or entire; not pinnatifid

7. Leaves mostly basal. Blades 4 to 12 inches long

Perennial, 8 to 32 inches high with mostly basal leaves. Leaves large, triangular ovate to heart-shaped, 4 to 12 inches long and 2 to 8 inches wide; entire or crenate and sparsely hispid. Heads large, showy, with 12 to 20 yellow ray flowers, ¾ to 1¼ inches long. Heads solitary or few at ends of mostly leafless stems. Achenes glabrous, without a pappus. In sandy soil from northern Los Angeles County northward. Blooms April to June.

Asteraceae
Balsamorhiza deltoidea Nutt. **Deltoid Balsamroot**

---

7. Leaves not mostly basal

8. Leaves opposite; at least the lower ones (Leaves alternate, page 148)

9. Leaves succulent. Plant of coastal salt marsh

Glabrous perennial; lower stem horizontal and creeping. Leaves sessile, oblanceolate; succulent; ¾ to 1½ inches long. Heads solitary. Involucral bracts rounded, in 2 to 3 series; pinkish above. Rays 6 to 10, narrow, less than ¼ inch long and somewhat inconspicuous. Common in coastal salt marsh. Blooms from May to October.

Asteraceae
Jaumea carnosa (Less.) A. Gray

Marsh Jaumea

---

9. Leaves not succulent

10. Leaves linear to filiform

11. Leaves filiform. Rays conspicuous, up to ⅜ inch long

Low, slender, unbranched annual 2 to 6 inches high in dry, clay soils; up to 10 inches tall in good soil. Leaves opposite filiform, entire, with a clasping base. Heads terminal with 6 to 13 yellow rays ¼ to ⅜ inch long. Abundant in early spring, covering hillsides and open grasslands, making them yellow. Blooms March to May.

Asteraceae
Lasthenia californica Lindl.
subsp. *californica*

California Goldfields

---

11. Leaves linear. Ray inconspicuous, hidden by the involucral bracts

Low annual, 4 to 16 inches high; villous and more or less glandular above. Leaves clasping; opposite below and alternate above. Leaves linear, up to 5 inches long. Flowering heads solitary on long peduncles. Involucres ⅝ to ¾ inch high. Ray flowers yellow, inconspicuous and barely longer than the green involucral bracts. Disk flowers 15 to 35. More noticeable than the flowers are the fruiting heads. The achenes are crowned with a pappus of silvery scales forming a head which slightly resembles a dandelion. Blow-wives is quite common in grassy fields with clay soil. Blooms from April to May.

Asteraceae
Achyrachaena mollis Schauer

Blow-Wives

---

10. Leaves lanceolate to ovate

11. Leaves lanceolate

12. Herbage rough hispid

Erect, perennial herb up to 5 feet tall. Leaves opposite; 1 to 5 inches long and ⅜ to 1¼ inches wide; lanceolate, with entire or slightly toothed margins and acuminate apex. Leaves more or less hispid with a sandpapery texture. Flowers solitary on long peduncles, forming an open inflorescence. Heads large and showy with yellow ray and disk flowers. Ray flowers ⅝ to 1 inch long. Achenes glabrous with a pappus of 2 awns which are readily deciduous. Involucral bracts are hispid. Common on dry hillsides from May to October.

Asteraceae
Helianthus gracilentus A. Gray

Slender Sunflower

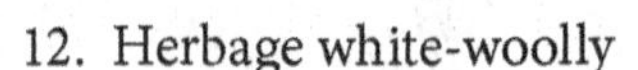

12. Herbage white-woolly

White-woolly annual 4 to 24 inches high; erect, with few branches. Leaves lanceolate, entire, or undulate dentate. Heads terminal, showy; about 1 to 2 inches across. Involucral bracts in 1 series; white-woolly with acute tips. Rays 8, bright yellow, ⅜ to ¾ inch long. Common plant in grassy areas and valleys. Blooms in spring from March to May.

Asteraceae
Monolopia lanceolata Nutt.

Common Hillside Daisy

---

11. Leaves ovate; at least the lower leaves

12. Herbage rough hispid. Disk reddish brown

Tall, stout annual 1 to 6 feet or more tall; openly branched. Herbage with numerous stiff hairs giving a scabrous or sandpapery texture. Leaves ovate to deltoid with a truncate base, 2 to 6 inches long; serrate and distinctly petiolate. Heads large and showy, 2 or more inches across; sticky pubescent. Involucral bracts ciliolate, slightly imbricated; ovate, with long acuminate tips. Rays golden yellow; disk reddish brown. Abundant weed of field and roadside. Grows to large size during wet years and covers entire fields during late summer. Blooms from February to October. Sunflower seeds are quite delicious, especially if roasted first.

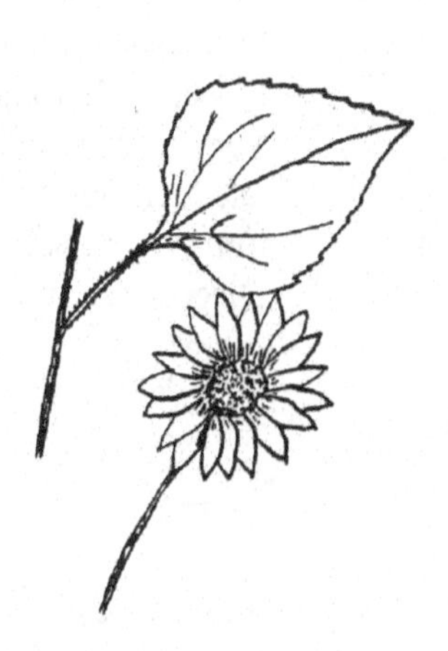

Asteraceae
Helianthus annuus L.

Common Sunflower

---

12. Herbage gray pubescent. Disk yellow

Erect annual 1 to 4 feet high. Leaves opposite below and alternate above. Lower leaves deltoid, sharply toothed. 1½ to 4 inches long. Upper leaves lanceolate. Stems and under surface of leaves grayish pubescent. Heads terminal on long peduncles; showy; 1 to 2 inches across. Ray flowers yellow or orange, about ½ inch long with a 3-toothed apex. Involucral bracts in 2 or 4 series; lanceolate with grayish pubescence. Achenes black, later becoming whitish-green; flat with wings on either side which end in an awn. Pappus of two short awns. Weed in fields occurring from Ventura to Riverside Counties. Blooms from May to December.

Asteraceae
Verbesina encelioides (Cav.) A. Gray
subsp. *exauriculata* (B.L. Rob. & Greenm.)
J.R. Coleman

Golden Crownbeard

---

8. Leaves alternate

9. Leaves linear to lanceolate (Leaves broader, page 151)

10. Ray flowers 5. Disk flowers 6

Erect herb 8 to 40 inches tall, simple below but widely (divaricately) branched above. Herbage hirsute to almost glabrous. Leaves mostly gone on lower stem at time of flowering. Upper leaves small, linear; not much fascicled. Heads clustered at ends of short branches; numerous. Rays 5; yellow, 3-lobed. Disk flowers 6. Involucral bracts, not much imbricated. Very common plant in dry fields and open hillsides in late summer. Blooms from May to August. Tarweed is found from Santa Barbara County southward to San Diego County.

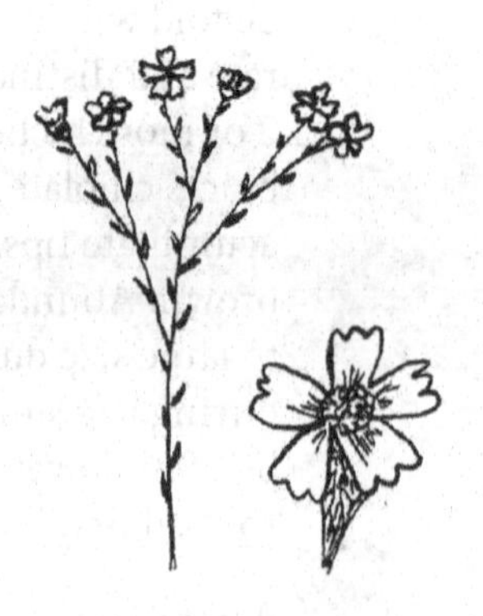

Asteraceae
Deinandra fasciculata (DC.) Greene

Fascicled Tarweed

10. Ray flowers more than 5. Disk flowers more than 6
  11. Leaves toothed
      See *Senecio californicus*, page 145.
  11. Leaves entire
    12. Herbage resinous; glandular
      13. Ray flowers inconspicuous; ⅛ to ¼ inch long

Fragrant and resinous herb 4 to 40 inches high with sessile, linear leaves. Leaves up to 4 inches long and ¼ inch wide, glandular, hairy. Heads in an open panicle with 8 to 12 yellow rays ⅛ to ¼ inch long and 15 to 35 disk flowers. Involucral bracts have acuminate tips and completely enclose the ray achenes. Achenes obovate and mottled. Pappus none. Rather abundant in open places. Blooms from April to August.

Asteraceae
Madia gracilis (Sm.) Applegate

Gumweed

---

      13. Ray flowers conspicuous; ¼ to ⅝ inch long

Glandular pubescent to villous herb 8 to 32 inches high; much and openly branched above. Upper stem very sticky, glandular. Lower leaves densely villous; lanceolate, ⅛ to ¼ inch wide; 2 to 4 inches long; sessile, entire. Upper leaves much reduced, linear and bract-like. Heads glandular hirsute; numerous and somewhat showy. Ray flowers 8 to 16; yellow; ¼ to ⅝ inch long. Disk flowers yellow or maroon with black anthers. Common plant on dry chaparral slopes. Blooms from June to August.

Asteraceae
Madia elegans D. Don

Common Madia

---

12. Herbage not glandular

13. Plant low, less than 1 foot high. Leaves less than 1½ inches long

Low, slender annual, 3 to 12 inches high with linear, glabrous leaves. Lower leaves about ⅜ to 1¼ inches long. Upper leaves shorter. Flowering heads yellow; ⅜ to 1 inch across. Ray flowers numerous, $^{3}/_{16}$ to ½ inch long. Disk flowers also numerous. *Pentachaeta* grows in dry, open, grassy places and may be confused with goldfields. The leaves are alternate, however, and the ray flowers are more numerous. It also blooms slightly later, April to July. Lyon's Pentachaeta is lightly pubescent and is found mostly near the coast in Los Angeles and Ventura Counties and Santa Catalina Island. Its distribution is very limited and is listed as **RARE** and **ENDANGERED**.

Asteraceae
Pentachaeta lyonii A. Gray **Lyon's Pentachaeta**

---

13. Plant tall, 1 to 7 feet high. Leaves 1½ to 4 inches long

Glabrous herb 2 to 7 feet high. Leaves linear-lanceolate; sessile, entire with acuminate apex; 1½ to 4 inches long; ⅛ to ⅜ inch wide. Heads yellow with 15 to 25 ray flowers and 7 to 14 disk flowers. Involucral bracts imbricated Heads about ⅛ inch high in broad or open, terminal clusters. Pappus of many white bristles. Found in moist areas such as wet meadows, stream banks, and fresh or salt water marshes. Blooms from July to November.

Asteraceae
Euthamia occidentalis Nutt. **Western Goldenrod**

---

9. Leaves broader, lanceolate to ovate

10. Leaves entire or mostly so

11. Plant white-woolly

See *Monolopia*, page 147.

11. Plant not white-woolly

12. Leaves 2 to 6 inches long

13. Leaves spathulate. Disk yellow. Flowers small, numerous

Perennial herb 8 inches to 3 feet high. Leaves alternate, obovate-spathulate; 2 to 5 inches long; ⅜ to 1½ inches wide. Lower leaves serrate; upper leaves entire and reduced. Leaves and stem with grayish pubescence. Heads yellow; $^{3}/_{16}$ inch high; 8 to 13 ray flowers and 4 to 12 disk flowers. Involucral bracts narrow with herbaceous tips. Achenes hispid with a pappus of many white bristles. Heads in a dense, elongate cluster. In dry or moist fields and in cleared areas. Blooms from July to October.

Asteraceae
Solidago velutina DC.
subsp. *californica* (Nutt.) Semple

California Goldenrod

---

13. Leaves deltoid-ovate. Disk reddish brown. Flowers large, few

See *Helianthus annuus*, page 147.

12. Leaves ⅜ to ¾ inch long

Erect perennial, 12 to 32 inches high. Leaves crowded, alternate, entire. Upper leaves sessile, ⅜ to ¾ inch long, lanceolate-ovate; grayish hirsute. Flowering heads numerous with both ray and disk flowers. Involucres $^{3}/_{16}$ inch high. Sessileflower goldenaster grows in dry, sandy areas away from the immediate coast. Blooms from July to November.

Asteraceae
Heterotheca sessiliflora (Nutt.) Shinners

Sessileflower Goldenaster

---

10. Leaves toothed; at least the lower leaves

11. Involucral bracts recurved; very sticky, gummy. Leaves prickle-tipped

Erect, glabrous plant 1½ to 4 feet high. Leaves lanceolate with a broad clasping base, sharply toothed. Heads 1 to 2 inches across, terminal at ends of branches. Involucral bracts imbricated with long, green, recurved tips; resinous, gummy. Rays 25 to 45, bright yellow. Common in chaparral, particularly along coast. Blooms March to September.

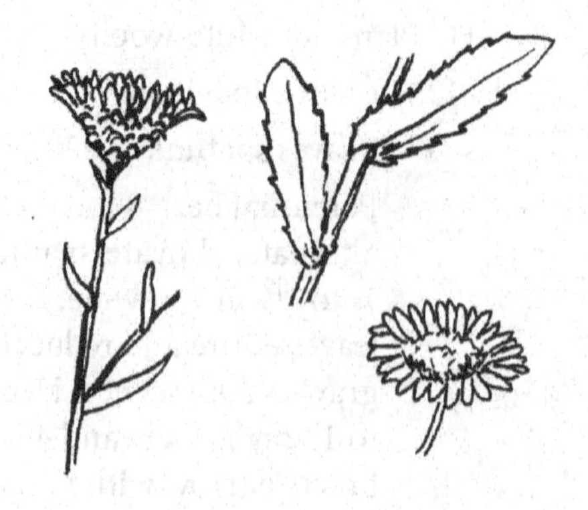

Asteraceae
Grindelia camporum Greene

Common Gumplant

---

11. Involucral bracts not recurved; not sticky or gummy

12. Ray flowers ⅝ to 1 inch long

13. Plant with sandpapery texture. Disk reddish brown

See *Helianthus annuus*, page 147.

13. Plant glabrous. Disk yellow

See *Venegasia*, page 143.

12. Ray flowers shorter

13. Ray flowers ⅜ to ⅝ inch long; 3-lobed

See *Verbesina*, page 148.

13. Ray flowers ⅜ inch or less; not 3-lobed

14. Ray flowers 25 to 35; curling upon drying. Plant weedy, densely hairy

Coarse, erect plant with stout stem, unbranched below; 1½ to 7 feet tall. Stem densely hirsute; glandular pubescent in inflorescence. Leaves ovate, serrate; thick and densely villous; ¾ to 3 inches long. Leaves sessile above; petiolate below. Heads numerous, about ¾ to 1 inch across; in a compound inflorescence. Rays 25 to 35, yellow; curling rapidly upon drying. Common weed of late summer, but can be found blooming most of the year.

Asteraceae
Heterotheca grandiflora Nutt.

Telegraph Weed

---

14. Ray flowers 8 to 13; not curling upon drying
   15. Heads small, numerous. Ray flowers ⅛ inch long
       See *Solidago velutina*, page 151.
   15. Heads larger; few. Ray flowers ⅜ inch long
       See *Senecio californicus*, page 145.

4. Ray flowers not yellow
 5. Ray flowers purple or blue
  6. Leaves linear. Herbage not white-woolly

Tall, erect, branched perennial 8 to 40 inches high, glabrous to slightly pubescent. Leaves linear to oblong, ¾ to 2¼ inches long. Heads numerous in an open inflorescence. Involucres about ¼ inch high. Involucral bracts imbricated. Rays numerous, 20 to 60; linear; lavender to blue; ¼ to ½ inch long. Disk flowers yellow. Common plant on grassy and brushy slopes of the chaparral. Blooms from May to July.

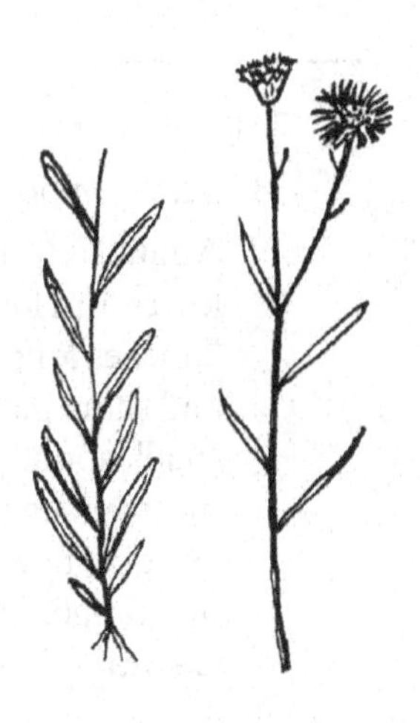

Asteraceae
Erigeron foliosus Nutt.

Leafy Fleabane

---

  6. Leaves lanceolate. Herbage white-woolly

Slender, erect perennial covered with white wool; 16 to 32 inches high. Leaves lanceolate, ¾ to 2½ inches long; becoming reduced above. Heads ¼ inch high. Involucral bracts much imbricated with green tips, recurving at maturity. Rays violet to purple. Disk flowers yellow. Common plant of late summer on dry, brushy chaparral slopes. Blooms from July to October.

Asteraceae
Corethrogyne filaginifolia (Hook. & Arn.) Nutt.

California Aster

---

5. Ray flowers white
   6. Leaves simple (Leaves pinnatifid dissected, page 155)
      7. Ray flowers showy; ¼ to ⅝ inch long

Erect annual 4 to 24 inches high. Leaves hispid, lower ones pinnately toothed or lobed; upper ones entire. Heads few, terminal with hispid involucral bracts. Rays white, showy, ¼ to ⅝ inch long with 3-lobed apex. Disk yellow. Blooms in spring from March to June. Occasional spring annual in dry, sandy soil.

Asteraceae
Layia glandulosa (Hook.) Hook. & Arn.

White Layia

---

7. Ray flowers short, inconspicuous; barely exceeding disk
   8. Leaves opposite. Ray flowers 4 or 5

Annual 8 to 18 inches high with opposite leaves and long inter-nodes. Leaves ovate, ¾ to 2 inches long; mostly toothed with acute apex and short petioles. Herbage glabrous. Heads small, about ⅛ inch high. Ray flowers white, inconspicuous and barely surpassing the yellow disk flowers. Pappus of a few scales. Weed in Orange, San Bernardino, and Los Angeles Counties. Flowers from May to November.

Asteraceae
Galinsoga parviflora Cav.
subsp. *parviflora*

Gallant Soldier

---

8. Leaves alternate
   9. Leaves ovate, palmately lobed and toothed. Rays ⅛ inch long

Annual with somewhat brittle, ascending stems 4 to 16 inches long. Leaves alternate, pubescent, ovate to cordate; palmately lobed; the lobes toothed. Leaves ⅜ to 1½ inches long on petioles about the same length; glandular pubescent. Ray flowers white, inconspicuous; 1/16 to ⅛ inch long. Disk flowers yellow. Involucral bracts in 2 series, with ciliate margins and tips. Occurs in the desert and in coastal sage scrub from Ventura to Los Angeles County. Flowers from February to June.

Asteraceae
Perityle emoryi Torr.

Emory's Rock Daisy

---

9. Leaves oblanceolate to linear, entire or slightly toothed. Rays filiform and very inconspicuous. Common weed in waste places

10. Involucral bracts glabrous to strigose

Tall, erect, leafy annual up to 6 feet tall; simple below, much branched above. Leaves numerous, glabrous to strigose, oblanceolate below, linear above; up to 4 inches long; entire or serrate. Lower leaves petiolate; upper ones sessile. Heads small, about 3⁄16 inch high; numerous, in a terminal panicle. Involucral bracts green and glabrous. Rays white, but inconspicuous, barely surpassing the disk. Pappus of many white capillary hairs. Common in late spring and summer in fields and waste places. Blooms from June to September.

Asteraceae
Erigeron canadensis L.

Horseweed

---

10. Involucral bracts soft hairy

Erect, branched annual ½ to 3 feet high, with gray-green, soft hairy herbage. Leaves oblanceolate below to linear above; serrate to entire. Heads small, about 3⁄16 inch high; numerous, in a terminal panicle. Involucral bracts green, hairy. Ray flowers present; white, but small and inconspicuous, barely visible above the pappus. Pappus of numerous whitish capillary hairs. Common weed of lawn, garden, and waste places. Blooms from June to August.

Asteraceae
Erigeron bonariensis L.

Flax-Leaved Horseweed

---

6. Leaves pinnatifid dissected

7. Flowers all disk. Outer flowers enlarged and ray-like

Stout annual 1 to 5 feet high with whitish pubescent herbage. Leaves pinnatifid dissected into linear lobes. Heads showy; about ⅜ inch high. Flowers white, all disk, but outer flowers enlarged and sometimes appearing ray-like. *Chaenactis artemisiaefolia* occurs in sandy soil, particularly near the coast. **Fleshy Pincushion**, *Chaenactis xantiana* A. Gray, occurs farther inland on the desert slopes such as around Mt. Pinos and base of the San Gabriel Mountains. *C. xantiana* is less than 20 inches tall and lacks the white, scaly pubescence.

Asteraceae
Chaenactis artemisiifolia (Harv. & A. Gray) A. Gray

White Pincushion

---

7. Flowers not all disk

8. Heads numerous, in a flat-topped inflorescence. Disk flowers white to pink

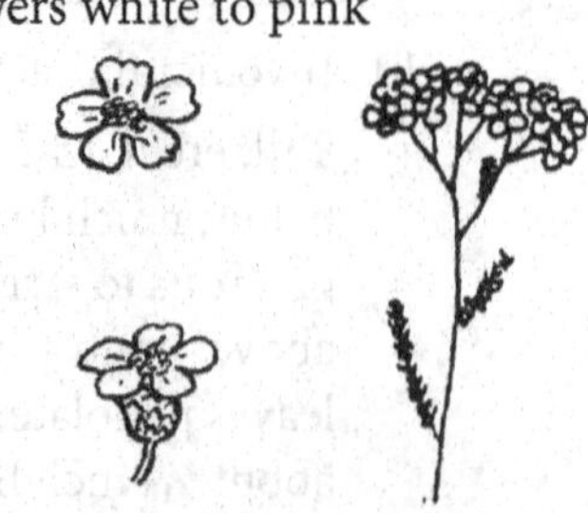

Perennial weed 1 to 3½ feet high. Leaves pinnately dissected into very fine fernlike divisions. Heads small and numerous in a flat-topped inflorescence. Involucral bracts soft pubescent and distinctly imbricated. Rays 3 to 8, white; about 1/16 inch long. Common weed of summer.

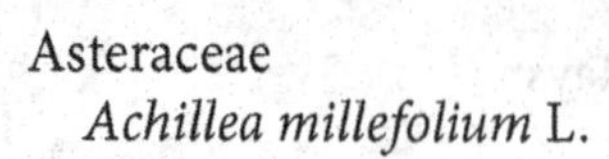

Asteraceae
Achillea millefolium L.

Yarrow

---

8. Heads solitary at branch tips. Disk flowers yellow

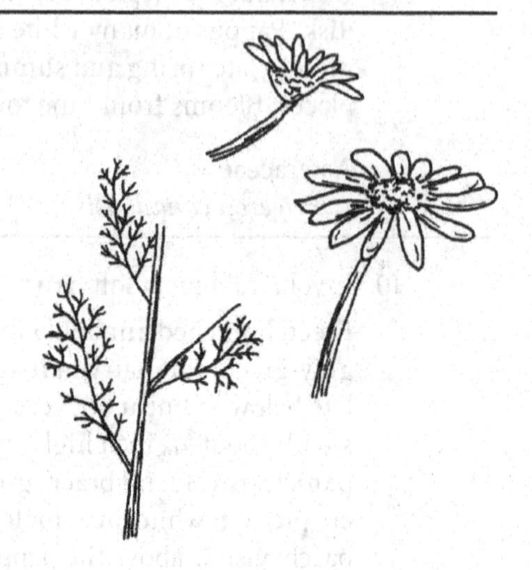

Ill-scented, annual weed 4 to 20 inches tall with almost glabrous herbage. Leaves ¾ to 2½ inches long; pinnately dissected into fine divisions. Heads numerous, solitary on ends of branches; ½ to ¾ inch across. Rays white. Common weed of waste places. Blooms from April to August.

Asteraceae
Anthemis cotula L.

Mayweed

---

2. Heads with only ray flowers. No disk flowers present

3. Ray flowers rose to lavender; 2-lipped. Juice not milky

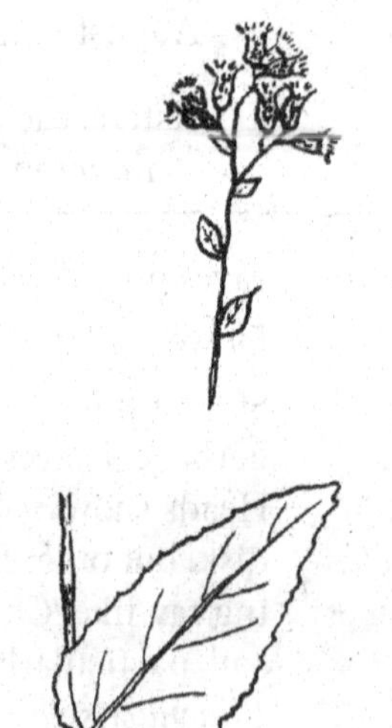

Branched perennial up to 3 feet high with scabrous stems and leaves. Leaves ovate, alternate, sessile, sharp or spinose toothed. Heads in a large, open panicle. Heads up to ½ inch high with many lanceolate, acuminate involucral bracts. Bracts sometimes reddish. Flowers rose to lavender; 2-lipped; the upper lip 2-lobed. Lower lip broad and 3-toothed at the apex. Pappus of numerous white, capillary hairs. Fairly common plant of wooded canyons and chaparral slopes. Blooms from June to August.

Asteraceae
Acourtia microcephala DC.

Sacapellote

---

3. Ray flowers not rose-lavender and 2-lipped. Juice milky

4. Flowers blue, violet, or pink

5. Flowers pale pink to lavender. Lower leaves oblong; sinuate; or remotely toothed

6. Leaves withered at flowering time. Heads less than ½ inch high

Erect, glabrous annual 1½ to 7 feet tall with stiff, erect branches above middle. Lower leaves sinuate, oblong, withered before flowering. Upper leaves small, linear. Heads almost sessile on leafless gray-green branches. Heads a little more than ¼ inch high; 4 to 15 flowered. Rays pale pink to lavender or almost white. Common in dry fields and hills. Blooms July to October.

Asteraceae
Stephanomeria virgata Benth. subsp. *virgata*

Rod Wire-Lettuce

---

6. Leaves not withered at flowering time. Heads ½ to ⅝ inch high

Perennial, 1 to 4 feet high; woolly when young. Leaves oblong-oblanceolate, remotely toothed or almost entire; sessile, with acute apex. Heads pink with 12 ray flowers ½ to ¾ inch long. Achenes with a pappus of plumose or feathery bristles which soon fall off. On rocky slopes and in canyons in the Santa Ana and San Bernardino Mountains and north. Blooms August to October.

Asteraceae
Stephanomeria cichoriacea A. Gray

Chicoryleaf Wire-Lettuce

---

5. Flowers bright blue. Lower leaves oblanceolate, pinnatifid

Erect, glabrous, branched perennial 1 to 3½ feet high; woody at base. Leaves mostly basal, oblanceolate, deeply toothed or pinnatifid, 4 to 8 inches long. Upper leaves few and reduced. Heads sessile on almost leafless branches, in axils of much reduced bract-like leaves. Involucral bracts oblong, sometimes reddish. Heads showy, up to 1½ inches across, with bright blue ray flowers. Occasional in waste places. Not abundant in Ventura County.

Asteraceae
Cichorium intybus L.

Chicory

---

4. Flowers yellow or white; sometimes veined with rose
  5. Leaves mostly basal (Leaves on main stem, page 160)
    6. Achenes (fruits) with a beak and a tuft of capillary bristles (pappus) on top

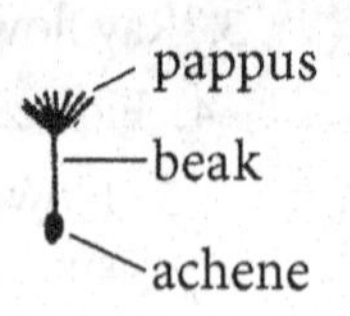

---

7. Leaves and involucral bracts pubescent

Robust perennial from a woody root crown. Scapes 6 to 24 inches high. Leaves all basal; lanceolate, 4 to 10 inches long; lobed or laciniate to entire; pubescent to lightly woolly. Heads erect, 1 to 1½ inches high. Inner involucral bracts linear lanceolate; outer ones ovate; densely pubescent. Rays bright yellow. Fruit an achene with a long beak about twice the length of the achene, topped with a pappus of numerous, white, capillary hairs. Occasional plant of the chaparral woodland. Blooms from May to July.

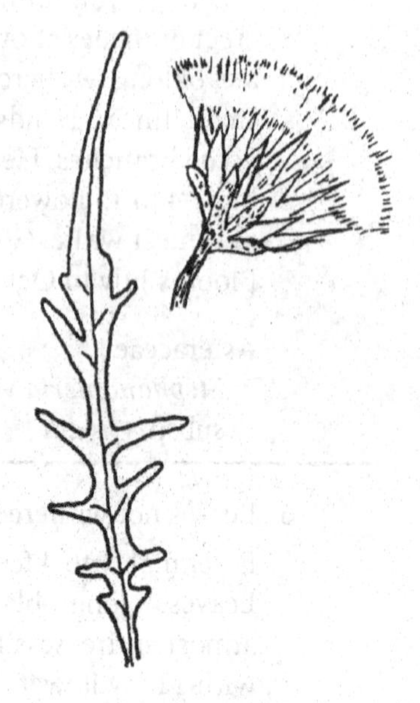

Asteraceae
Agoseris grandiflora (Nutt.) Greene — **Mountain Dandelion**

---

7. Leaves and involucral bracts glabrous
  8. Common weed of lawns. Heads large, ¾ to 2 inches across. All achenes beaked

Glabrous herb with leaves all basal and more or less flat on the soil. Leaves deeply pinnatifid or toothed, 2 to 12 inches long. Heads terminal on a reddish, hollow scape 2 to 8 inches high. Heads ¾ to 2 inches across; rays bright yellow. Achenes long beaked with a tuft of capillary hairs at the summit. Very common weed of lawn and waste places. Blooms most of the year. Dandelion leaves may be eaten either cooked or raw. They may tend to be a bit bitter, however. The roots are also supposed to have some medicinal value and can be used as a tonic or as a mild laxative.

Asteraceae
Taraxacum officinale F.H. Wigg. — **Common Dandelion**

---

8. Weed. Heads ¼ inch across or less. Only inner achenes beaked

Glabrous annual with mostly basal leaves. Leaves lobed; oblong to spathulate, up to 4 inches long. Heads ¼ inch or more across with yellow ray flowers that scarcely exceed the involucral bracts. Achenes glabrous with a pappus of plumose bristles ¾ inch long. Weed. Blooms March to June.

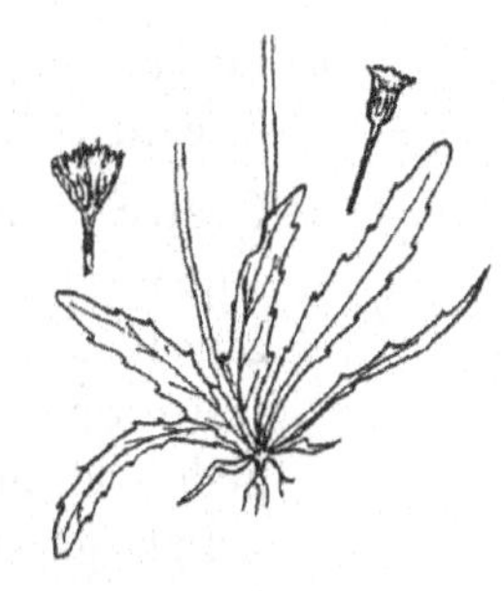

Asteraceae
Hypochaeris glabra L.

Smooth Cat's-Ear

---

6. Achenes not beaked.

7. Pappus of 2 to 5 flat scales with a bristle tip. Leaves all basal

Low, glabrous annual, 4 to 24 inches high with basal leaves. Leaves entire or pinnatifid, 2 to 8 inches long. Heads terminal, nodding, with up to 100 yellow or white ligules. Involucres ½ inch high. Pappus scales tapered to a barbed awn-like tip. Similar are: **Elegant Silverpuffs,** *Microseris elegans* A. Gray, with ligules yellow to orange; involucres only ¼ inch high and pappus awn tips not barbed; and **Lindley's Silverpuffs,** *Uropappus lindleyi* (DC.) Nutt., with heads that are erect, rather than nodding. Found in grassy areas. Blooms March to June.

Asteraceae
Microseris douglasii (DC.) Sch. Bip.

Douglas' Microseris

---

7. Pappus of numerous bristles. Leaves not all basal

Glabrous, branched annual, 2 to 14 inches high. Basal leaves toothed or pinnately lobed. Cauline leaves much reduced. Heads, ¼ inch high; 20 to 60 flowered. Ligules yellow, ¼ inch long. Pappus of 12 to 32 bristles. Found in burned or disturbed areas of the chaparral. Blooms April to June.

Asteraceae
Malacothrix clevelandii A. Gray

Annual Malacothrix

---

5. Leaves on main stem
  6. Flowers yellow (Flowers white, page 162)
    7. Leaves simple
      8. Leaves rough bristly with short spines on upper surface. Involucral bracts large and leaflike, spine tipped

Coarse, bristly annual with erect, branched stems 1 to 3 feet long. Leaves alternate, sessile, ovatelanceolate, 2 to 8 inches long, rough bristly, with short spines on upper surface. Heads terminal on short peduncles or at ends of branches; ½ to 1 inch high with large, showy yellow rays. Involucral bracts in 2 series; the outer ones almost leaflike, spine-tipped, and spreading; the inner ones, lanceolate and much smaller. Common weed in waste places. Blooms June to December.

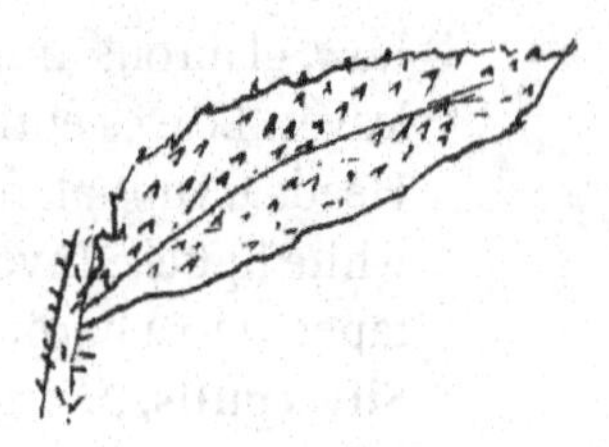

Asteraceae
Helminthotheca echioides (L.) Holub

Bristly Ox-Tongue

---

8. Leaves sparsely hispid. Involucral bracts narrow

Spreading annual 4 to 13 inches high with hispid stem and leaves. Basal leaves toothed or lobed with a winged petiole. Cauline leaves oblanceolate, with serrate margin, 1 to 4 inches long. Heads on long naked peduncles. Ray flowers yellowish with a purple tip, ¼ to ⅜ inch long. Pappus of thin scales $^{1}/_{16}$ inch long and some slightly longer bristles. Involucral bracts of 1 series, ¼ inch long, enfolding the marginal achenes. Achenes about $^{3}/_{16}$ inch long; without a beak. Weed. Blooms April to May.

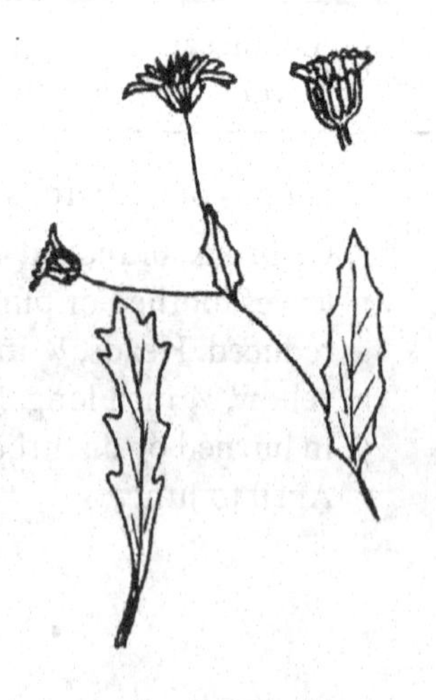

Asteraceae
Hedypnois cretica (L.) Dum. Cours.

Crete Weed

---

7. Leaves pinnately dissected or parted
   8. Plants with sharp prickles on stem and on under surface of leaves along midrib

Tall, prickly annual 2 to 5 feet high. Prickles abundant in lower part of stem. Leaves pinnatifid or deeply toothed; clasping, 1½ to 6 inches long, spinose-toothed and with sharp spines on under surface along midrib. Heads numerous in an open panicle. Flowers pale yellow but not particularly showy. Heads ⅜ inch high with imbricated involucral bracts. Pappus of numerous soft, white, capillary hairs. Common weed in fields and waste places. Blooms May to September.

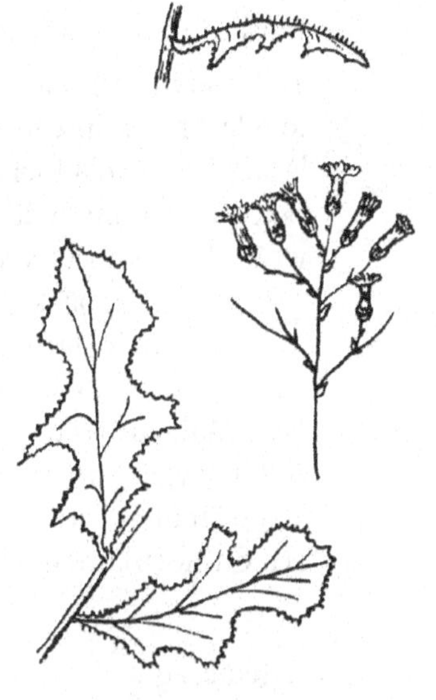

Asteraceae
Lactuca serriola L.

Prickly Lettuce

---

8. Plants without sharp prickles on stem and leaves
   9. Leaves with pointed auricles at base

Rather weak stemmed glabrous annual, erect and 1½ to 8 feet high. Leaves 4 to 8 inches long with a large deltoid, terminal lobe and winged petioles; sharply toothed. Leaves clasping with a pointed, auricled base. Heads terminal, yellow, about ½ inch high. Pappus of numerous, white, capillary hairs. Very common weed in gardens and waste places. Blooms most of the year.

Asteraceae
Sonchus oleraceus L.

Common Sow Thistle

---

9. Leaves with rounded auricles at base

Very similar to the above except that the auricles are rounded rather than pointed. Also very common weed in lawns, gardens, and waste places. Blooms most of the year.

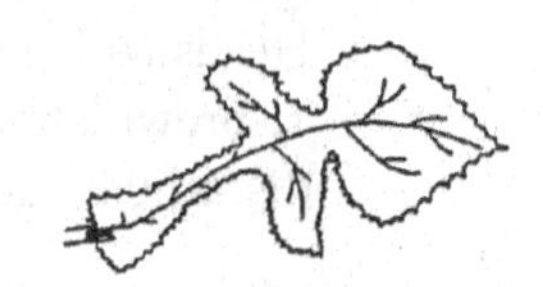

Asteraceae
Sonchus asper (L.) Hill
subsp. *asper*

Prickly Sow Thistle

---

6. Flowers white; may be veined with rose
   7. Rays white with a rose stripe on back

Tall, leafy perennial, somewhat woody at the base; 1 to 5 feet high, spreading and much branched. Leaves lanceolate, 1 to 4 inches long, toothed or entire. Heads ⅜ to ⅝ inch high with lanceolate involucral bracts which have scarious margins. Ray flowers white with rose stripes on back. Very common chaparral plant on dry soil; particularly abundant near the coast. Blooms from March to September. Also present is the **Thin-Leaf Malacothrix**, *Malacothrix saxatilis* (Nutt.) Torr. & A. Gray var. *tenuifolia* (Nutt.) A. Gray, with pinnately leaves. It grows inland in the chaparral along the southern California coast and on the Channel Islands.

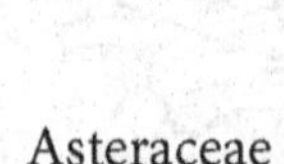

Asteraceae
Malacothrix saxatilis (Nutt.) Torr. & A. Gray

Rock Malacothrix

---

7. Rays white, not veined with rose
   8. Heads sessile on leafless branches. Achenes not beaked

Erect, glabrous herb 1½ to 7 feet tall with stiff, erect branches above the middle. Lower leaves sinuate, oblong, withered before flowering. Upper leaves small and linear. Heads almost sessile on the leafless gray-green branches. Heads a little more than ¼ inch high; inner involucral bracts green, lanceolate. Rays pale pink to lavender or almost white. Common in late summer in dry fields and hills. Blooms from July to October.

Asteraceae
Stephanomeria virgata Benth.
subsp. *virgata*

Rod Wire-Lettuce

---

8. Heads terminal on ends of branches, forming an open panicle. Achenes beaked

Glabrous or pubescent annual 8 to 60 inches high. Leaves entire to pinnatifid; 2 to 8 inches long. Lower leaves with petioles; upper leaves auriculate clasping. Heads with 15 to 30 white ray flowers. Pappus of 10 to 15 brownish, capillary bristles. In burned areas and disturbed places. Blooms from April to July.

Asteraceae
Rafinesquia californica Nutt.

California Chicory

---

1. Flowers all disk flowers; no ray flowers present
  2. Shrubs, woody at the base (Herbs, page 166)
    3. Leaves pinnatifid dissected into linear lobes. Herbage with distinct mint or sage odor

Grayish, shrub 2 to 5 feet tall with a distinct woody base. Leaves numerous, pinnately dissected into filiform divisions; often fascicled. Heads small, ovoid, ⅛ to 1/16 inch across; in an elongate, raceme-like inflorescence. Common plant on dry chaparral slopes. Blooms from August to December.

Asteraceae
Artemisia californica Less.

California Sagebrush

---

3. Leaves simple; toothed or entire
  4. Leaves minute and scale-like. Broom-like shrub

Broom-like shrub 3 to 6 feet tall with green, glabrous stems and yellow flowers. Leaves alternate, entire; much reduced and scale-like. Heads numerous in terminal racemes. Involucral bracts imbricated and scale like. Disk flowers yellow. Common in washes and sandy or gravelly places such as dry stream beds. Blooms August to October.

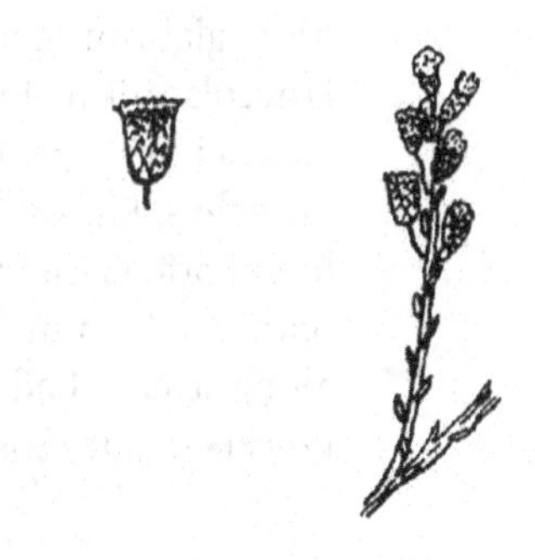

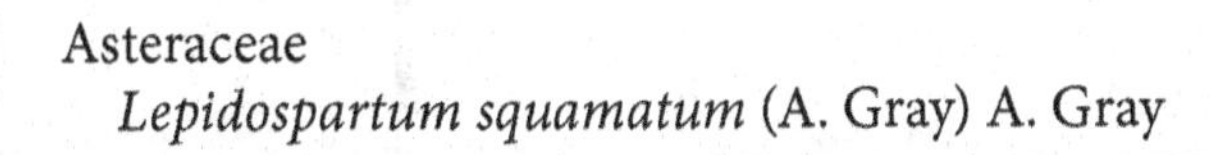

Asteraceae
Lepidospartum squamatum (A. Gray) A. Gray

Scalebroom

---

4. Leaves not minute and scale-like
  5. Herbage silvery silky. Flowers purplish

Willow-like shrub, 3 to 12 feet high, with silvery, silky herbage. Leaves somewhat crowded, entire, lanceolate, ⅜ to 1½ inches long and ⅛ to ¼ inch wide. Leaves alternate, sessile with an acute apex. Heads in terminal clusters. Flowers all disk flowers; purplish. Involucres about ¼ inch high. Arrow-weed is quite common in wet places and along rivers. Blooms March to July.

Asteraceae
Pluchea sericea (Nutt.) Coville

Arrow-Weed

---

5. Herbage not silvery silky. Flowers not purplish
   6. Leaves narrow linear to lanceolate
      7. Leaves linear to filiform

Erect shrub 2 to 9 feet high with glabrous, resinous, glandular-dotted stem and leaves. Leaves linear to filiform, 1½ to 2½ inches long; crowded on stems. Heads with 18 to 23 yellow disk flowers, in rounded panicles. Involucres 3/16 inch high. Involucral bracts, much imbricated, thin, chaffy. Golden fleece grows on the dry foothills of the chaparral from Ventura County north. Blooms from August to November.

Asteraceae
Ericameria arborescens (A. Gray) Greene — **Golden-Fleece**

---

7. Leaves lanceolate; willow-like

Willow-like shrub or small tree, 6 to 12 feet high; glabrous or slightly resinous. Leaves linear lanceolate, 1 to 4 inches long; dentate or entire. Heads in compact, terminal clusters at ends of branches. Heads 3/16 inch high; whitish. Involucral bracts imbricated and slightly papery or scarious. Common in moist places, particularly along stream beds. Blooms March to July, but flowers occur most of the year.

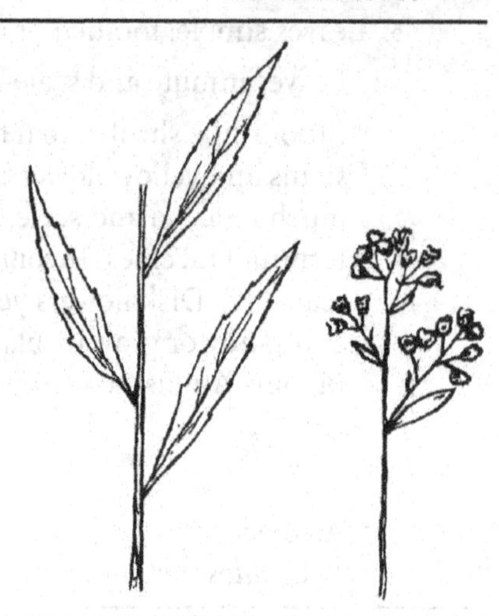

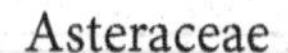
Asteraceae
Baccharis salicifolia (Ruiz & Pav.) Pers. — **Mule Fat**

---

6. Leaves broader; ovate, spatulate, or heart-shaped
   7. Leaves heart-shaped or triangular; apex acute

Much branched shrub 1½ to 4 feet high with rough, pubescent herbage, particularly when young. Leaves alternate, simple, deltoid or heart-shaped; ½ to 1½ inches long; serrate. Flowers all disk; white. Heads in clusters at ends of branches. Involucral bracts green, glabrous. Common shrub on dry slopes. Blooms in late summer; August to October.

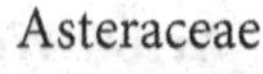
Asteraceae
Brickellia californica (Torr. & A. Gray) A. Gray — **California Brickellbush**

---

7. Leaves obovate; apex obtuse
  8. Low shrub, 1 to 3½ feet high. Leaves sharp toothed. Disk flowers yellow
    9. Leaves ⅜ to ¾ inch wide. Chaparral shrub

Much branched, erect and rigid shrub 1 to 3½ feet high, glabrous or slightly woolly above. Leaves numerous, leathery, clasping; obovate, sharply serrate with spinose teeth. Leaves mostly ½ to 1½ inches long. Flowers all disk, yellow, in heads ⅜ to ½ inch high. Involucral bracts much imbricated with green tips. Pappus of many red-brown, capillary hairs. Common shrub of coastal ranges and chaparral. Blooms July to October.

Asteraceae
Hazardia squarrosa (Hook. & Arn.) Greene — **Saw-Toothed Goldenbush**

---

9. Leaves ⅛ to ⅜ inch wide. Shrub of coastal dunes

Very leafy, erect shrub 1 to 3½ feet high; resinous to glandular pubescent. Leaves small, spatulate; mostly ¼ to 1¼ inches long; spinose toothed above middle; usually with axillary fascicles. Heads numerous in a dense terminal cluster. Heads ¼ inch high with imbricated, green-tipped involucral bracts. Flowers all disk, yellow. Pappus of many brownish capillary hairs. Common along coast. Blooms April to December.

Asteraceae
Isocoma menziesii (Hook. & Arn.) G.L. Nesom — **Coastal Goldenbush**

---

8. Tall shrub, 4 to 9 feet high. Leaves not sharp toothed. Disk flowers white.

Erect, much branched shrub 4 to 9 feet high with glabrous or resinous herbage. Leaves small, numerous; obovate, obtuse with 5 to 9 coarse teeth; mostly ½ to 1½ inches long. Heads ⅛ inch high, numerous, in small axillary and terminal clusters. Involucral bracts whitish, papery, and imbricated. Common along coast; in coastal ranges. Blooms August to December.

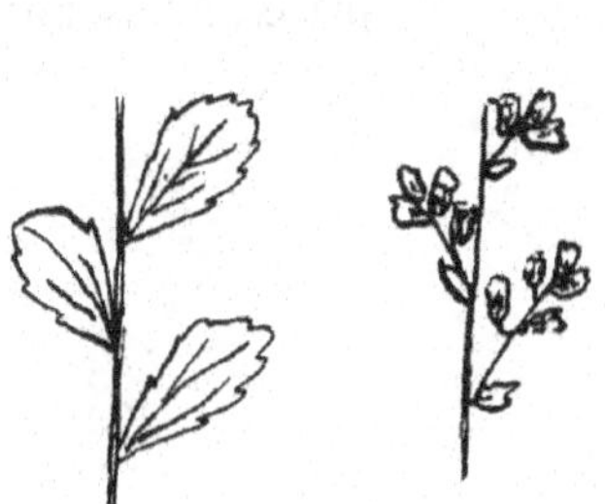

Asteraceae
Baccharis pilularis DC. — **Coyote Brush**

---

2. Herbs, not woody at the base

   3. Plant thistle-like with spine-tipped leaves and/or bracts (Plant not thistle-like, page 168)

      4. Involucral bracts spine-tipped. Leaves not spine-tipped

Annual 1 to 2½ feet high with some cottony pubescence, particularly in the inflorescence. Leaves linear lanceolate, entire above, lobed below; sessile with a decurrent base. Heads globose, averaging ½ inch across. Involucral bracts ending in a long spine. Flowers bright yellow. Common plant along roadsides, in fields, and in dry disturbed areas. Blooms from May to October. Two species are likely to be found in the area and may be distinguished by the length of the involucral spines. *Centaurea melitensis*, has spines ⅜ inch long or less. Spines on **Yellow Star-Thistle,** *Centaurea solstitialis* L., are longer; ⅜ to ¾ inch long.

Asteraceae
Centaurea melitensis L. **Tocalote**

---

      4. Leaves and involucral bracts spinose

         5. Leaves with white mottled veins. Flowers red-violet

Tall, stout herb 3 to 6 feet tall. Leaves large, green with white mottling on the veins; sinuate dentate with spinose teeth. Leaves clasping with an auriculate base, but leaves not decurrent down the stem. Heads large and globose, 1 to 2 inches across. Involucral bracts broad at the base and spine-tipped. Flowers red-violet to purple. Common plant in waste places, fields, and along roadsides. Blooms from May to July.

Asteraceae
Silybum marianum (L.) Gaertn. **Milk Thistle**

---

5. Leaves without white mottling on veins
  6. Heads large, 1 or more inches across
    7. Leaves, stem, and heads not densely white-woolly

Tall, stout herb 2 to 4 feet high. Leaves pubescent to slightly tomentose beneath; deeply pinnately lobed with stout prickly teeth. Leaves sessile with a prickly decurrent base on the main stem. Heads globose, 1 to 2 inches across, with numerous imbricated linear spine-tipped involucral bracts. Bracts spreading at maturity; the lower ones somewhat recurved. Flowers lavender to purple. Common chaparral plant and weed in waste places. Blooms from June to September.

Asteraceae
Cirsium vulgare (Savi) Ten.

Bull Thistle

---

7. Leaves, stem, and heads densely white-woolly

Tall, stout herb mostly 2 to 4 feet high, covered with a dense white tomentum. Leaves narrow lanceolate, pinnately lobed with long yellowish spine teeth. Leaves sessile and clasping but not decurrent on the main stem. Heads large, globose, 1½ to 2½ or more inches across. Involucral bracts linear lanceolate, rigid; spine-tipped, covered with white wool. Flowers dark crimson to red-purple, barely exceeding the involucral bracts. Common in coastal scrub, chaparral, and in disturbed areas. **California Thistle**, *Cirsium occidentale* (Nutt.) Jeps. var. *californicum* (A. Gray) D.J. Keil & C.E. Turner, differs in that has lighter colored flowers, lavender to whitish, which also well exceed the involucral bracts.

Asteraceae
Cirsium occidentale (Nutt.) Jeps.
var. *occidentale*

Cobwebby Thistle

---

6. Heads smaller, about ½ inch across

Annual with prickly stem and leaves; 1 to 6 feet tall. Leaves are pinnatifid; 4 to 5 inches long and 3 inches wide, with spine-tipped teeth; greenish and slightly woolly above and densely white-woolly beneath. Leaves extend down the stem forming a spinose wing. Heads are rose-purple, about ½ inch wide and occur in clusters of 2 to 5 at ends of stem. Involucral bracts are spine-tipped and covered with wool at base. Pappus consists of grayish bristles. Common in disturbed areas. Blooms from March to June.

Asteraceae
Carduus pycnocephalus L.

Italian Thistle

---

3. Plants not thistle-like. Involucral bracts not spinose
4. Plant of beach, dune, or salt marsh
5. Heads purple. Plant of salt marsh

Tall, erect annual to perennial, 1 to 4 feet high, ill-scented, with brownish pubescent to almost woolly stems. Leaves lanceolate, dentate, 2 to 4 inches long; densely pubescent, particularly on under surface. Heads in a large terminal inflorescence. Flowers purple, all disk. Involucral bracts purple; slightly papery. Occasionally found in salt marshes along the coast. Blooms from July to November.

Asteraceae
Pluchea odorata (L.) Cass.

Saltmarsh Fleabane

---

5. Heads not purple
6. Heads yellow, button-like. Leaves fleshy, sheathing

Low, spreading perennial with glabrous and somewhat fleshy stems. Leaves alternate deeply toothed or pinnatifid; ¼ to 2¾ inches long with scarious sheaths. Heads bright yellow, terminal, and button-like, composed of numerous compact disk flowers. Very common plant in moist, marshy areas, including salt marshes. Blooms from March to December.

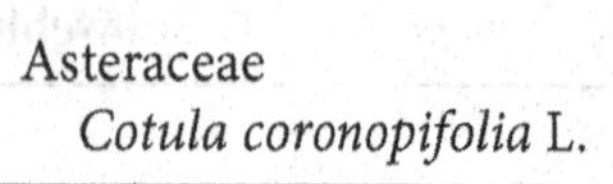

Asteraceae
Cotula coronopifolia L.

Brass-Buttons

---

6. Heads yellowish or greenish, inconspicuous. Leaves not as above

7. Prostrate perennial, forming mats on the sand. Leaves silvery pubescent. Fruit a bur

Prostrate or flat lying perennial forming mats on the sand 3 to 9 feet across and 6 to 12 inches high. Leaves silvery pubescent, pinnatifid into narrow, oblong segments. Heads unisexual in terminal spikes. Male heads with a bowl-shaped involucre, somewhat pendant. Female heads below the male, clustered in the leaf axils; becoming a sharp, prickly bur in fruit. Common on sand dunes along the coast. Blooms from March to September.

Asteraceae
Ambrosia chamissonis (Less.) Greene

Beach Bur-Sage

---

7. Erect annual. Leaves not silvery pubescent. Fruit not a bur. Flowers yellowish

Erect, annual herb 4 to 16 inches high with a sweet odor. Leaves alternate, linear; entire or pinnately divided into 3 to 5 linear lobes. Heads small in a panicle or cyme. Flowers all disk; 10 to 30; yellowish; inconspicuous. Achenes with a pappus of irregular scales. Along coast on beaches, dunes, and in salt marshes. Blooms March to June.

Asteraceae
Amblyopappus pusillus Hook. & Arn.

Dwarf Coastweed

---

4. Plant not of beach, dune, or salt marsh

5. Leaves pinnatifid or compound (Leaves simple, page 173)

6. Leaves pinnate compound into 3 to 5 ovate leaflets. Fruit with awns

Branched, slender-stemmed annual 1 to 5 feet high. Leaves pinnate compound with 3 to 5, ovate, serrate leaflets. Heads terminal; yellowish; disk flowers only. Involucral bracts ovate, green in center with a scarious, hyaline margin. Fruits with persistent awns, resembling ticks. Common weed in lowlands. Blooms from May to November.

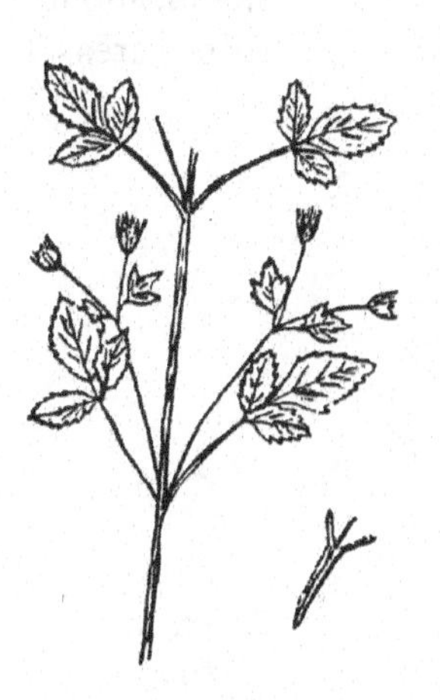

Asteraceae
Bidens pilosa L.

Common Beggar-Ticks

---

6. Leaves pinnatifid; not as above
  7. Fruit a prickly bur
    8. Annual. Spines on bur straight

Erect annual 4 to 28 inches high with slightly hispid stems. Leaves are ¾ to 2½ inches long; two times pinnatifid into oblong divisions. Heads are small and greenish or yellowish, clustered in the axils of leaves. Male heads are numerous; ⅛ inch across with a green, lobed or cap-shaped involucre. Female heads have a rudimentary corolla and develop into a spiny bur about ¼ inch long with straight spines. Pappus none. Common weed of sandy places from Fresno to San Diego County. Flowering from August to November.

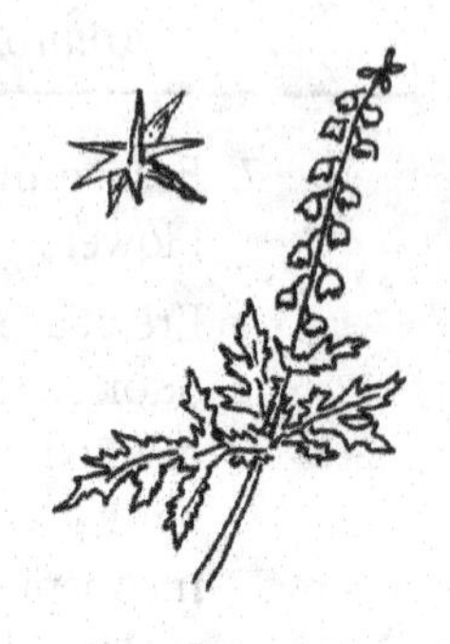

Asteraceae
Ambrosia acanthicarpa Hook. **Annual Bur-Sage**

---

8. Perennial. Spines on bur hooked

Erect perennial with grayish-green, hispid stems 1 to 4 feet high. Leaves are 2 times pinnatifid into linear oblong segments. Heads are small and greenish. Male heads are about ⅛ inch across with a cap-shaped or lobed involucre. Female heads form a bur with hooked spines; about ⅛ inch across. Common in dry, disturbed areas. Blooms from May until November.

Asteraceae
Ambrosia confertiflora DC. **Weakleaf Bur Ragweed**

---

7. Fruit not a prickly bur
   8. Flowering heads showy, ½ to 1½ inches across. Outer disk flowers ray-like
      9. Heads yellow
         See *Chaenactis glabriuscula*, page 144.
      9. Heads white
         See *Chaenactis artemisiifolia*, page 155.
   8. Flowering heads less than ½ inch across
      9. Plant tall, 1½ to 4 feet high. Heads small, greenish
         10. Common weed of late summer. Male heads nodding, with a cup-shaped involucre; in a terminal raceme

Rather strong-smelling, pubescent to hirsute perennial 1½ to 4 feet high. Leaves pinnatifid, 1½ to 4½ inches long, green to silvery pubescent. Male heads small, about ⅛ inch across, in a dense, terminal raceme. Heads nodding with a cup-shaped involucre. Female heads below the male in the upper leaf axils; 1-flowered, forming a tuberculate fruit; not spiny. Very common weed of late summer on roadsides, waste places, and open fields. Blooms from July to November.

Asteraceae
Ambrosia psilostachya DC.

Western Ragweed

---

10. Occasional weed. Heads globular, in a many-flowered panicle

Erect, glabrous annual or biennial, 12 to 32 inches high. Leaves once or twice pinnatifid into linear lobes. Heads globose, greenish, in a many-flowered panicle. Heads small; involucres about ⅛ inch high. Biennial wormword occurs as an occasional weed and blooms from August to October.

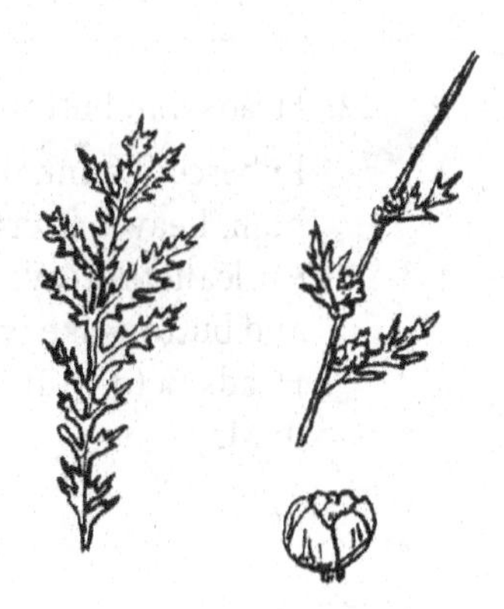

Asteraceae
Artemisia biennis Willd.

Biennial Wormword

---

9. Plant low, less than 1½ feet high

10. Plant of moist, marshy places. Leaves succulent. Heads ⅜ inch across

See *Cotula*, pages 168 and 172.

10. Weed of lawn and waste places. Leaves not succulent. Heads less than ⅜ inch across

11. Heads conical, with odor of pineapple when crushed

Low annual, erect and glabrous; 4 to 12 inches high. Leaves pinnatifid into linear divisions. Heads 1 to several, cone-shaped, sometimes hidden among the leaves. Flowers all disk, yellowish green. Involucral bracts at base of head slightly imbricated and with scarious margins. Very common weed of waste places. Blooms from May to August.

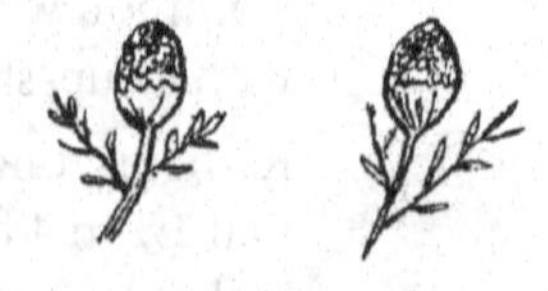

Asteraceae
Matricaria discoidea DC.

Pineapple Weed

---

11. Heads not conical and without odor of pineapple

12. Heads cylindrical. Herbage glabrous

Low, glabrous annual 4 to 20 inches high. Leaves pinnatifid into rounded toothed lobes. Upper leaves clasping. Heads several; cylindrical; about ¼ inch high. Flowers all disk; yellow. Involucral bracts black tipped; linear, in one series. Minute black-tipped bracts at base. Pappus of numerous capillary white hairs. Common weed of lawn, garden, and waste places. Blooms most of the year.

Asteraceae
Senecio vulgaris L.

Common Groundsel

---

12. Heads flat, button-like. Herbage pubescent

Pubescent annual with a strong scent; 1 to 8 inches high. Leaves alternate, pinnate compound with linear leaflets. Leaves ⅜ to 1¼ inches long. Heads small and button-like with numerous yellow disk flowers. Heads ⅛ to $^{3}/_{16}$ inch across. Weed. Flowers January to May.

Asteraceae
Cotula australis (Spreng.) Hook. f.

Australian Cotula

---

5. Leaves simple: toothed, lobed, or entire

6. Plant with 3-pronged spines on stem

Much branched, bushy annual 1 to 3½ feet high. Stems bearing numerous, stout, yellow, 3-pronged spines in leaf axils. Leaves green above; silvery beneath; lanceolate, entire, or 3 to 5 lobed; 1 to 3 inches across. Flowers in unisexual heads; in terminal and axillary clusters. Fruit a bur. Common weed in fields and waste places. Blooms from July to October.

Asteraceae
Xanthium spinosum L.

Spiny Cocklebur

---

6. Plant not spiny

7. Herbage white-woolly. Heads sometimes immersed in wool
(Herbage not white-woolly, page 177)

8. Low annuals, less than 12 inches high. Heads usually completely immersed in wool
(Plants taller, page 175)

9. Plant of moist or marshy areas

Low, erect annual, simple or branched; 2 to 12 inches high, densely white-tomentose. Leaves spathulate, ¼ to 1¼ inches long; upper leaves bract-like, subtending the dense, compact inflorescence; often surpassing it. Heads about ⅛ inch high, densely white-woolly at base; involucral bracts papery white. Common in moist or marshy areas. Blooms May to October.

Asteraceae
Gnaphalium palustre Nutt.

Lowland Cudweed

---

9. Plant of dry area

10. Heads globose, well surpassed by subtending leaves

11. Stem simple; mostly unbranched. Involucral bracts with a central green spot. Heads in axillary and terminal clusters

Low, floccose, woolly annual with a simple, slender, erect stem 2 to 14 inches high, but mostly 3 to 4 inches high. Leaves linear, acuminate; ¼ to ⅝ inch long. Heads terminal, clustered; often immersed in wool. Involucral bracts about 5; ovate, scarious; not imbricated. Common in dry, open places. Blooms April to June.

Asteraceae
Micropus californicus Fisch. & C.A. Mey. **Q-Tips**

---

11. Stem branched from base or above. Involucral bracts without a central green spot. Heads in terminal clusters

Erect annual 2 to 7 inches high with grayish, tomentose herbage. Leaves alternate, narrow oblong or spathulate; ⅛ to ¼ inch long; entire with obtuse apex. Heads clustered at tip of stem, subtended by tomentose, leafy bracts. Woolly bracts with a brownish, membranous wing surround the marginal flowers. Involucral bracts 5. On dry slopes and burns. Blooms March to May.

Asteraceae
Stylocline gnaphaloides Nutt. **Everlasting Neststraw**

---

10. Heads elongate, not much surpassed by subtending leaves

Erect annual, 2 to 14 inches high with white-woolly herbage. Leaves alternate, oblanceolate-linear to spathulate, entire; ⅜ to ¾ inch long. Heads in axillary and terminal clusters. Not surpassed by leaves. Outer flowers enclosed in a woolly bract. Central flowers rosy. In dry, open places and burned areas. Blooms March to June.

Asteraceae
Logfia filaginoides (Hook. & Arn.) Morefield **California Cottonrose**

---

8. Plants tall, 1 to 4 feet high. Heads often woolly but not completely immersed in wool. Involucral bracts white, papery. Herbage often sweet scented

9. Leaves green above (Leaves white-tomentose above, page 176)

10. Leaves white, tomentose beneath at maturity

11. Leaves of main stem linear. Involucral bracts chalky white

Perennial 1 to 2 feet high with tomentose stems. Leaves linear, ¾ to 3 inches long; green above; white tomentose beneath. Involucral bracts chalky, white, papery, ovate; ¼ inch long. Inhabitant of sandy areas in the chaparral. Blooms August to September.

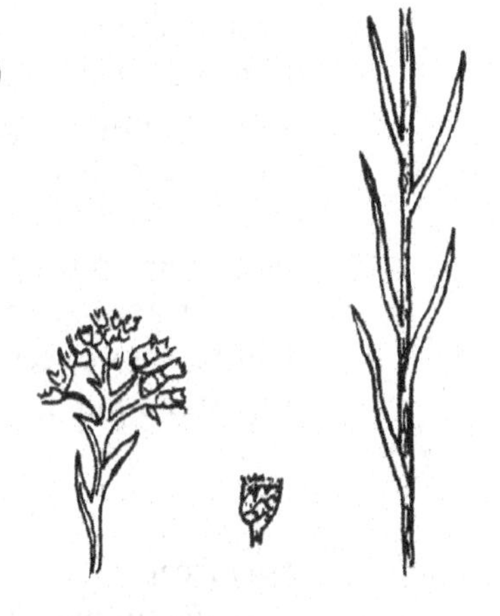

Asteraceae
Pseudognaphalium leucocephalum (A. Gray) Anderb.

White Rabbit-Tobacco

---

11. Leaves of main stem lanceolate. Involucral bracts silky, shiny

Perennial with white-woolly stems 1 to 3 feet high. Leaves lanceolate, ¾ to 2¾ inches long; green, glabrous above; densely, white-tomentose beneath with crisped margin; clasping by an auriculate base. Heads in dense, terminal clusters. Involucral bracts about ¼ inch long; ovate, shiny, lustrous. Common in dry open areas of chaparral. Blooms January to May.

Asteraceae
Pseudognaphalium biolettii Anderb.

Two-Tone Everlasting

---

10. Leaves green beneath at maturity

11. Involucral bracts pure white

12. Leaves lanceolate. Involucral bracts rounded; apex blunt

Stout biennial 1 to 3 feet high with green, glandular herbage and a distinct sweet curry scent. Lower leaves lanceolate, 1½ to 4 inches long; reduced above. Heads in a large terminal inflorescence. Involucral bracts white papery; about ¼ inch long; obtuse or rounded. Common in open places of the chaparral. Blooms January to July.

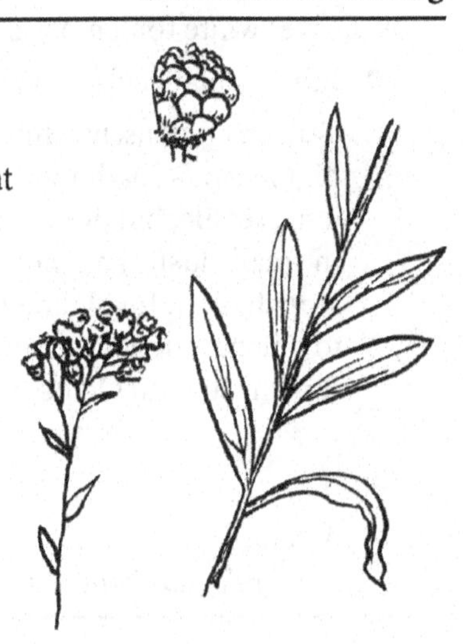

Asteraceae
Pseudognaphalium californicum (DC.) Anderb.

California Everlasting

---

12. Leaves lanceolate to linear; revolute. Bracts ovate. Stems rusty in age. Perennial; distinctly woody below

Erect perennial, distinctly woody at base. Stems 1 to 4 feet high. Herbage green with a slight amount of wool, becoming slightly rusty in age. Leaves lanceolate to linear-lanceolate, ¾ to 3 inches long; usually narrower and shorter above; sessile, entire, and more or less revolute. Involucral bracts pearly white, ovate. Heads in dense, terminal clusters. Herbage with a strong sweet curry scent. Blooms from June to August. Found in the San Bernardino Mountains.

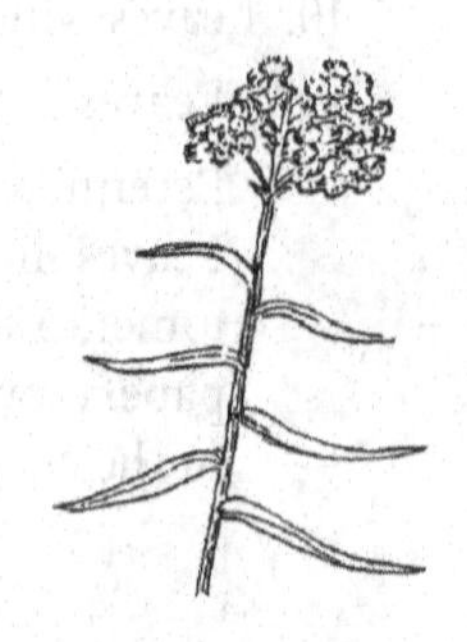

Asteraceae
Anaphalis margaritacea (L.) Benth. **Pearly Everlasting**

---

11. Involucral bracts pinkish

Tall, erect biennial 2 to 4 feet high; glandular and very strongly scented. Herbage green and glabrous except when young. Leaves linear lanceolate, ¾ to 2½ inches long. Heads in a large terminal inflorescence. Heads 3/16 inch high. Involucral bracts pinkish. Common plant in open areas, particularly along coast. Not usually found much inland. Blooms from March to September.

Asteraceae
Pseudognaphalium ramosissimum (Nutt.) Anderb. **Pink Everlasting**

---

9. Leaves white tomentose on both surfaces at maturity
10. Leaves oblanceolate; not decurrent on stem

Tall, erect, densely white-woolly biennial 1½ to 3½ feet high. Leaves spathulate or oblanceolate; ¾ to 2 inches long; sessile, but not decurrent on main stem. Heads in small clusters at ends of the many branches. Heads ¼ inch high. Involucral bracts papery white, ovate to oblong. Inhabitant of dry slopes and open places. Blooms June to October.

Asteraceae
Pseudognaphalium microcephalum (Nutt.) Anderb. **Feltleaf/White Everlasting**

---

10. Leaves linear-lanceolate; decurrent

Simple, erect perennial; 1½ to 3 feet high with densely white-woolly herbage. Leaves linear lanceolate, ¾ to 3¾ inches long; lower leaves wider. Heads in clusters at ends of branches ¼ inch high. Involucral bracts white, papery, sometimes slightly yellowish. Inhabitant of dry areas in chaparral. Blooms from July to November.

Asteraceae
Pseudognaphalium beneolens (Davidson) Anderb. **Fragrant Everlasting**

---

7. Herbage and heads not white-woolly. Involucral bracts not white, papery

8. Leaves triangular ovate. Fruit a bur

Coarse annual 1½ to 4 feet high with short, stiff appressed hairs giving a scabrous or sandpapery texture. Leaves alternate, large, deltoid with a truncate base; 1 to 4½ inches long and equally wide. Petioles as long as leaf blades. Leaves dentate or serrate. Female heads ovoid, becoming a bur ¾ to 1½ inches long; with numerous stout prickles and 2 short incurved beaks at top. Common weed in waste places. Blooms from July to October.

Asteraceae
Xanthium strumarium L. **Cocklebur**

---

8. Leaves linear, lanceolate to spatulate; not triangular ovate. Fruit not a bur
   9. Plant low, spreading, succulent. Flowers yellow, button-like
      See *Cotula*, pages 168 and 172.
   9. Plant erect; not succulent. Flowers not as above
      10. Heads solitary, ⅝ to ¾ inch high on long peduncles
          See *Achyrachaena*, page 146.
      10. Heads smaller, in dense, terminal clusters
          11. Leaves silvery beneath

Tall, erect perennial 1½ to 10 feet high, simple or branched. Leaves lanceolate, entire or few-lobed; 2½ to 6 inches long; dark green above; grayish white tomentose beneath. Heads small, turbinate, about ⅛ inch high; in a dense terminal, somewhat leafy inflorescence. Involucral bracts grayish. Common plant of low places. Blooms from June to October.

Asteraceae
Artemisia douglasiana Besser

Mugwort

---

11. Leaves green beneath
    12. Heads slightly elongate. Pappus of numerous capillary hairs, visible in fruit, giving a fuzzy appearance. Lower leaves toothed or entire
        See *Erigeron*, pages 153 and 155.
    12. Heads globular. No capillary hairs visible. Lower leaves often 3-cleft

Perennial herb ½ to 5 feet high with glabrous to villous stems and herbage. Leaves are linear; entire or 3-cleft below; 1 to 3 inches long. Heads are small and greenish, forming a panicle. Ray flowers are present but these are quite small and inconspicuous. Similar to the above is **Biennial Wormword,** *Artemisia biennis* Willd., except that leaves are 2 or 3 times pinnate into linear toothed lobes. Flowers occur in clusters in the axils of the leaves. Both species occur in waste places, but *A. dracunculus* is found more in dry areas, while *A. biennis* is in moist places. Blooms August to October.

Asteraceae
Artemisia dracunculus L.

Tarragon

SECTION IV
VINES, PARASITES, AND CLIMBING PLANTS

1. Plant body threadlike; yellow-orange; waxy

Stems orange, hair-like, waxy; twining over other vegetation, sometimes forming dense, colorful, tangled masses. Leaves reduced to scales and not conspicuous. Flowers small and white, waxy, occurring in clusters. Corolla and calyx white; 5 cleft. Very common parasite of the chaparral, blooming in late spring and summer often forming colorful masses on hillsides. Several species occur in the area and are typically host species specific:

Chaparral Dodder
Convolvulaceae
Cuscuta californica Hook. & Arn.
Common and parasitic on sage, buckwheat, deerweed, and goldenbush.

Salt Dodder
Convolvulaceae
Cuscuta salina Engelm.
Occurs on plants of salt marshes, particularly on marsh jaumea and pickleweed.

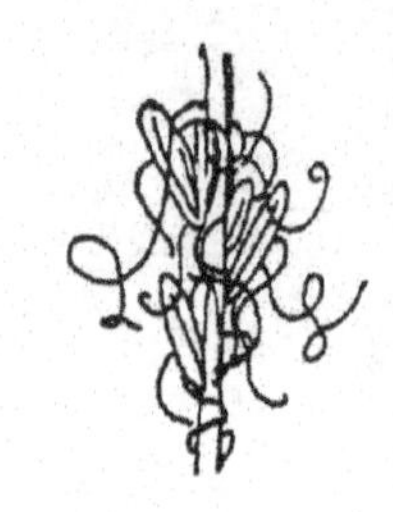

Canyon Dodder
Convolvulaceae
Cuscuta subinclusa Durand & Hilg.
Common and parasitic on sugar bush, gooseberry, monkey flower, *Encelia*, and nightshade.

Dodder or Witch's Hair

---

1. Plant body not threadlike and orange-yellow
2. Plant green, parasitic on trees. Leaves fleshy-ovate

Stem woody, brittle, penetrating into host. Leaves entire, thick, ovate with obtuse apex. Flowers unisexual, dioecious, inconspicuous. Petals absent. Fruit a white berry $\frac{3}{16}$ inch in diameter. Mistletoe blooms in late fall or winter and can be seen as a dense, straggly cluster among the upper branches of the host. **Big Leaf Mistletoe**, *P. s.* subsp. *macrophyllum*, is parasitic primarily on sycamore, but also found on popular, willow, and walnut with leaves that are generally glabrous. **Oak Mistletoe**, *P. s.* subsp. *tomentosum*, is commonly found on oaks with densely villous leaves. Other taxonomic considerations for mistletoe are *P. leucarpum* subsp. *macrophyllum* and *P. leucarpum* subsp. *tomentosum*.

Big Leaf Mistletoe
Viscaceae
Phoradendron serotinum (Raf.) M.C. Johnst. subsp. *macrophyllum* (Engelm.) Kuijt

Oak Mistletoe
Viscaceae
Phoradendron serotinum (Raf.) M.C. Johnst. subsp. *tomentosum* (DC.) Kuijt

Mistletoe

2. Plant not parasitic; merely climbing over other plants; vine-like

3. Flowers regular (Flowers irregular, page 183)

4. Fruit large, globose to oblong, 2 to 5 inches long

5. Flowers small, white. Fruit prickly

Climbing, trailing plant with tendrils. Leaves heart-shaped; palmately 5 to 7 lobed; 2 to 4 inches across. Flowers unisexual; small, in axils of leaves. Female flowers with a 5-lobed, white corolla. Ovary inferior. Fruit oblong, green and prickly, 3 to 5 inches long and 2 to 3½ inches across. Common plant of early spring along roadsides and dry areas of the chaparral. Blooms from January to June. Although a cut specimen smells exactly like cucumber, the fruit is extremely bitter and unpleasant tasting.

Cucurbitaceae
Marah macrocarpa (Greene) Greene

Wild Cucumber or Chilicothe

---

5. Flowers large, yellow-orange. Fruit a smooth gourd

Coarse scabrous perennial, ill-scented; trailing extensively along the ground and often extending 6 to 12 feet. Leaves large, 6 to 12 inches long, triangular; somewhat erect and partially folded along midrib. Flowers, large, yellow; campanulate; about 4 inches long; solitary in leaf axils. Corolla pubescent; ovary inferior. Fruit a smooth gourd; green with white stripes or mottling. Common weed in sandy soil. Easily recognized by the unpleasant odor emitted even without picking or crushing the leaves. Blooms in summer from June to August. The fleshy root may be cut into pieces and used as soap.

Cucurbitaceae
Cucurbita foetidissima Kunth

Buffalo Gourd or Calabazilla

---

4. Fruit not large, globose to oblong
  5. Leaves simple. Stamens 4 or 5
    6. Leaves alternate
      7. Flowers ¾ to 1 inch long

Prostrate, twining plant with triangular or sagittate leaves. Leaves ½ to 1½ inches long with rounded apex. Flowers in leaf axils on short, slender peduncles. Corolla, ¾ to 1 inch long 5-lobed; funnelform; pleated; white to pale pink. Calyx 5-lobed, ⅛ inch long. Flowers subtended by 2 small linear bracts about ¼ or more inches below the flower. Very common, troublesome weed of lawn and garden. Blooms from May to October.

Convolvulaceae
Convolvulus arvensis L.

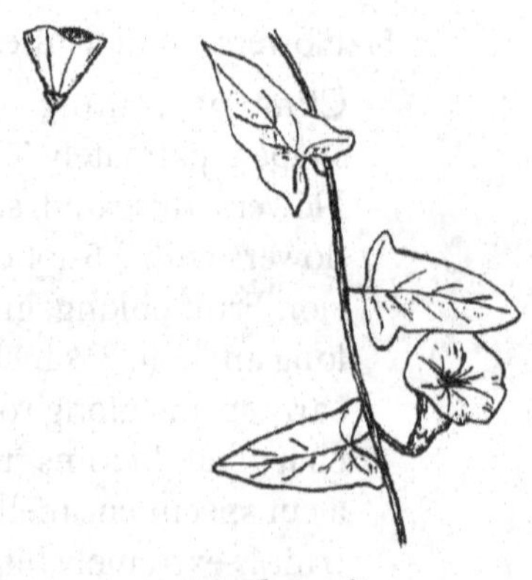

Bindweed

---

7. Flowers 1 to 1¾ inches long

Perennial with twining stems 3 to 12 feet long. Leaves ovate to deltoid-lanceolate; ¾ to 2 inches long. Flowers solitary in axils, on peduncles 1 to 4 inches long. Corolla funnelform; 5-lobed; 1 to 1¾ inches long; white with purple stripes and becoming pink in age. Sepals lance-ovate. Bracts subtending the flowers just beneath the calyx; membranous and calyx-like; oval to ovate; ½ to ⅝ inch long. Common on dry slopes of southern California. Blooms from March to August.

Convolvulaceae
Calystegia macrostegia (Greene) Brummitt

Wild Morning-Glory

---

6. Leaves whorled
  7. Leaves in whorls of 6 to 8

See *Galium aparine*, page 59.

  7. Leaves in whorls of 4

See *Galium nuttallii*, page 59.

5. Leaves compound. Stamens numerous (more than 10)
  6. Leaflets 5 to 7. Flowers small, ½ to 1 inch across

Woody, twining vine with opposite, compound leaves. Leaflets 5 to 7, lanceolate ovate, entire, few-toothed, or 3-lobed. Flowers small, white, in dense clusters. Petals none; sepals 4 or 5, petal-like. Stamens numerous. Fruit a fluffy, powderpuff-like ball of feathery tailed achenes. Chaparral plant of coast ranges. Blooms from March to August.

Ranunculaceae
Clematis ligusticifolia Nutt.

Western Virgin's Bower

---

6. Leaflets 3. Flowers larger, ¾ to 1½ inches across

Climbing, woody perennial with pubescent stems. Leaves opposite, compound. Leaflets 3; ovate; broadly toothed or lobed. Flowers showy with 4 to 5 white, pubescent petal-like sepals ½ to 1 inch long. Stamens numerous. Fruit a tailed achene densely clustered in a powderpuff-like ball. Common plant in chaparral canyons. Blooms from March to June.

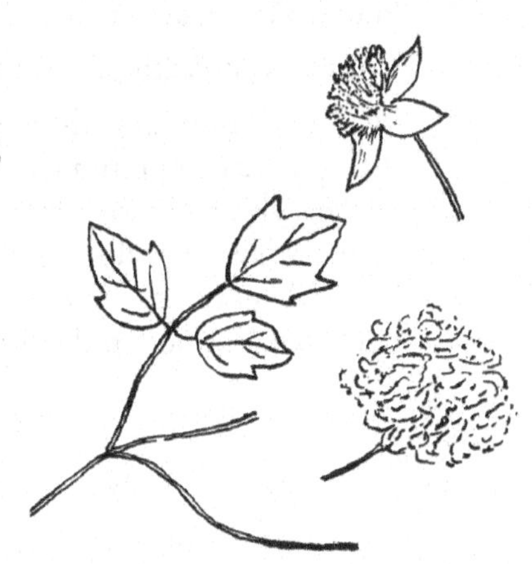

Ranunculaceae
Clematis lasiantha Nutt.

Chaparral Clematis

---

3. Flowers irregular; papilionaceous
4. Flowers sessile, single or paired in leaf axils

Slightly pubescent, twining annual with stems up to 3 feet long. Leaves compound with 8 to 16 leaflets, obovate with a truncate or emarginate apex. Flowers in leaf axils; subsessile; 1 to 2 flowered. Corolla violet-purple, ¾ to 1 inch long. Weed in waste places but not particularly common. Blooms April to June.

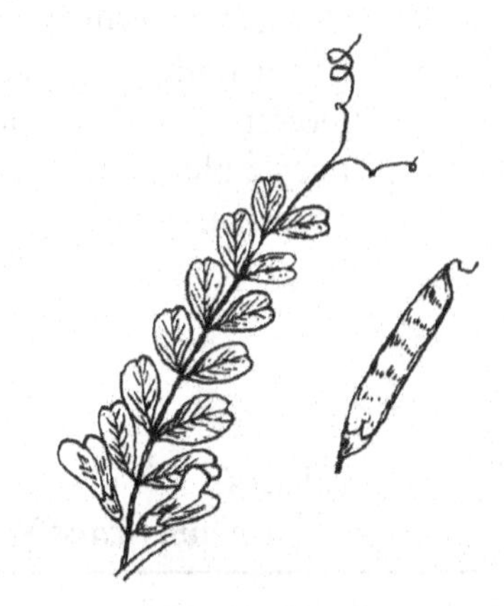

Fabaceae
Vicia sativa L.

Spring Vetch

---

4. Flowers at the ends of peduncles
5. Leaflets 4 to 16
6. Flowers 1 to 3; small, $\frac{3}{16}$ inch long

Slender annual, 18 to 24 inches high with climbing, villous stems and well developed tendrils. Leaves with rounded, obtuse or emarginate apex. Flowers 1 or 2, on slender peduncles that are shorter than the leaves. Corolla papilionaceous, white to purplish, $\frac{3}{16}$ inch long. Fruit a glabrous pod, ¾ to 1¼ inches long. Blooms from April to June.

Fabaceae
Vicia ludoviciana Torr. & A. Gray

Deerpea Vetch

---

6. Flowers 4 to 15; larger; ⅝ to 1 inch long, in terminal racemes

7. Corolla white with pink or lavender markings. Common

Twining perennial with stems terminating in a tendril. Leaves compound with 8 to 12 linear to ovate leaflets with a minute mucro tip. Flowers in axillary racemes of 5 to 15 flowers. Corolla white with pink or lavender markings; ⅝ to ¾ inch long. Fairly common in chaparral woodland. Blooms from April to June.

Fabaceae
Lathyrus vestitus Nutt.

Canyon Pea

---

7. Corolla purple-blue. Not common

Twining or trailing perennial. Leaves compound with 8 to 16 oblong leaflets with a rounded or obtuse apex. Flowers in axillary racemes of 3 to 9 flowers. Corolla purple-blue; ¾ to 1 inch long. Occasional on protected chaparral slopes. Blooms from April to June.

Fabaceae
Vicia americana Willd.

American Vetch

---

5. Leaflets 12 to 24

Twining annual or biennial with villous stems 2 to 4 feet long. Leaves pinnate compound with 6 to 12 pairs of linear to oblong, entire leaflets. Leaflets mucronate; ¼ to 1 inch long. Stipules narrow, entire; ⅜ inch long. Flowers violet and white, ⅝ inch long in a one-sided raceme. Calyx villous; ¼ inch long. Fruit a glabrous, oblong legume ¾ to 1¼ inch long. Becoming naturalized in waste places. Blooms from April to July. Similar to this is **Winter Vetch,** *Vicia villosa* Roth subsp. *varia* (Host) Corb., which is mostly glabrous rather than villous and has flowers that are purple-violet; ⅜ to ⅝ inch long.

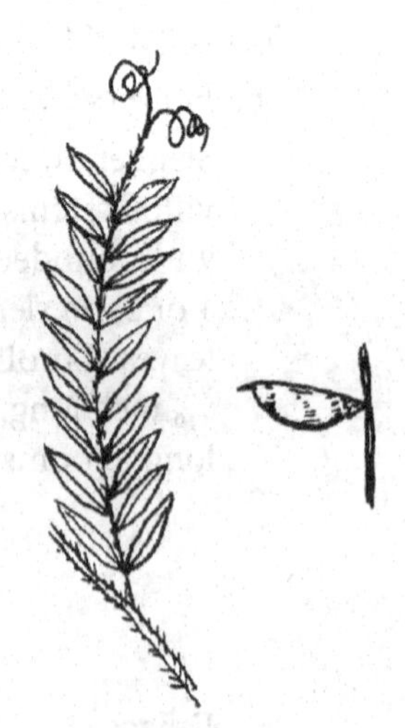

Fabaceae
Vicia villosa Roth subsp. *villosa*

Hairy Vetch

SECTION V
CACTI

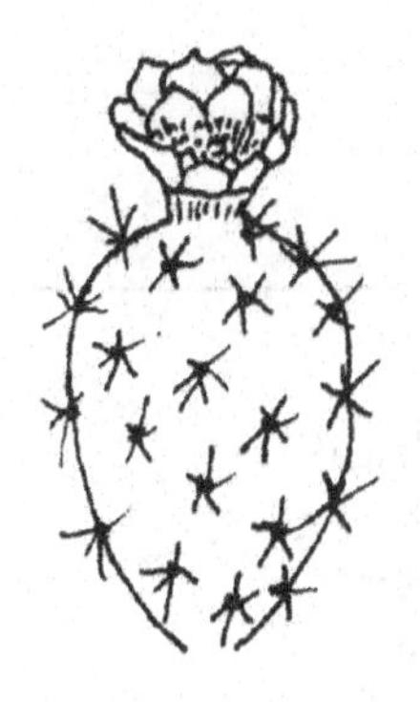

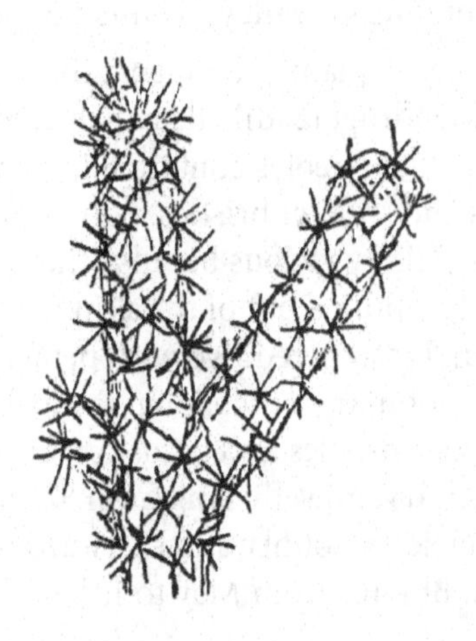

1. Joints cylindrical **"Cholla"**
   2. Coastal plant. Fruits prolific, growing one upon another

Erect, spiny plant, 2 to 6 feet high, with many spreading cylindrical branches. Terminal joints 1 to 6 inches long and easily detached so that they may be found lying on the ground. These joints can then root and form a new plant. Areoles each contain many small bristles (glochids) and 4 to 12 rusty yellow spines ⅜ to ¾ inch long. Spines barbed and very painful if they get into the skin. Flowers rose to dark red, with many imbricated waxy petals about ⅝ inch long. Fruits subglobose, ¾ to 1¼ inches long, fleshy, and prolific, bearing flowers and new fruits. Forms thickets on arid slopes near and along coast; never found inland. From Ventura County south to San Diego County.

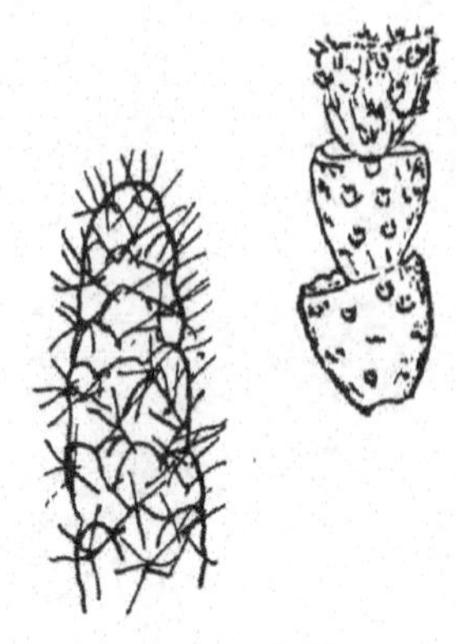

Cactaceae
Cylindropuntia prolifera (Engelm.) F.M. Knuth

Coast Cholla

---

2. Plant of interior valleys. Fruits not prolific

Erect, fleshy plant, 2 to 8 feet tall. Branches cylindrical and ascending; readily detachable. Joints up to 12 inches long. Areoles contain 6 to 20 yellow to brown spines and smaller bristles. Spines ⅜ to 1 inch long and especially vicious because they are barbed so they cannot be pulled out of the skin without being very painful. Flowers reddish with petals about 1 inch long. Fruits globular, ⅝ inch long. Fruits have areoles with numerous bristles and occasionally a few spines. Common on dry, gravelly fans from Santa Barbara County to San Diego County and eastward to the edge of the desert. Blooms from May to June.

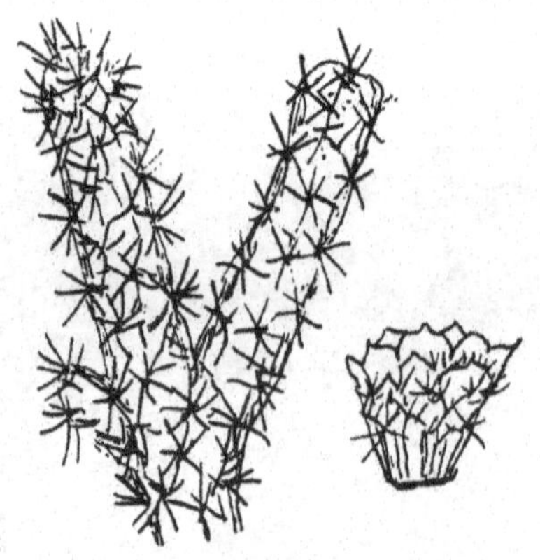

Cactaceae
Cylindropuntia californica (Torr. & A. Gray) F.M. Knuth var. *parkeri* (J.M. Coult.) Pinkava

Valley Cholla

---

1. Joints flattened, pad-like **"Prickly Pear"**

   2. Plant low, less than 12 inches high. Spines lacking

      Low, spreading cactus with gray-green or purplish, flat pads that are obovate; 5 to 10 inches long, and transversely wrinkled. Pads do not contain spines, but the areoles contain numerous glochids or barbed bristles which are painful if they get into the skin. Flowers showy, bright rose to magenta, up to 3 inches across. Perianth segments up to 1½ inches long. Fruits dry when ripe. Beavertail cactus is quite common in the desert, but is not common in the chaparral. It may be found in drier sites through the area.

Cactaceae
Opuntia basilaris Engelm. & J.M. Bigelow

Beavertail

---

   2. Plant taller, 1 to 15 feet high. Spines mostly present

      3. Plant cultivated; tree-like, 9 to 15 feet tall

         Large, shrubby, tree-like cactus 9 to 15 feet high. Pads are fleshy, rounded to elongate, and nearly spineless, 12 to 24 inches long and 8 to 16 inches wide. Spines, if present, are white to tan; straight; up to 1½ inches long. Flowers yellow or orange and quite showy. Fruit yellow to reddish; fleshy and edible; 2 to 3 inches long. This is probably the best of the edible cacti fruits. Although no spines occur on the fruits, there are numerous glochids, so care must be taken to remove them before eating. Young pads may also be eaten if the outer skin is removed. Mission Prickly-Pear was cultivated in Mexico for its fruits when the Spanish arrived and was introduced as one of the mission cacti. Blooms from May to June.

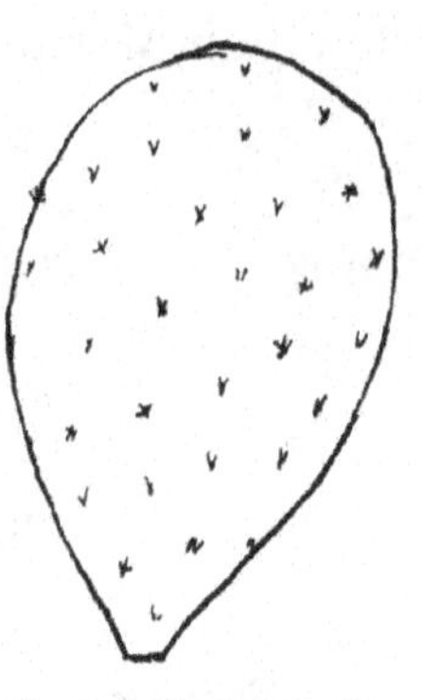

Cactaceae
Opuntia ficus-indica (L.) Mill.

Mission Prickly-Pear

---

3. Native cactus; smaller, less than 9 feet high

4. Plant with tree-like trunk at base; 3 to 9 feet tall. Pads circular in outline

Large cactus, 6 or more feet high with tree-like trunk up to 1 foot high. Pads fleshy, circular in outline, and 6 to 8 inches long. Spines, 8 to 16 per areole, translucent yellow, some of which have a downward curvature. Flowers yellow, showy, with 1½ inch long petals. Fruits globular and lacking a narrowed base; red, fleshy; 1 to 1¼ inches long. Fruits edible but care must be taken to remove the glochids before eating. Chaparral prickly pear grows in sandy soil below 500 feet. It blooms from April to May.

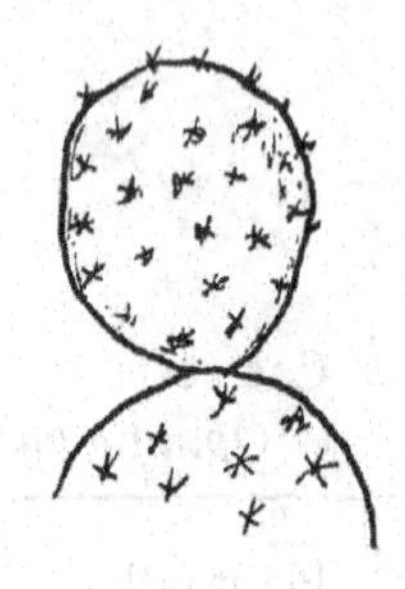

Cactaceae
Opuntia oricola Philbrick

Chaparral Prickly-Pear

---

4. Plant without trunk-like base; spreading, 1 to 5 feet high. Pads obovate

Somewhat erect or spreading plant up to 5 feet high, without a trunk-like base. Pads narrowly obovate, 4½ to 9 inches long and 2½ to 4 inches wide. Spines 1¼ to 1½ inches long; 5 to 11 per areole. Flowers pale yellow; showy; with many imbricated waxy petals 1½ to 2 inches long. Stamens many. Fruits pear-shaped, reddish to purple; 1½ to 2 inches long; fleshy and edible. Coast prickly pear grows at low elevations near the coast from Santa Barbara to Lower California. It blooms from May to June.

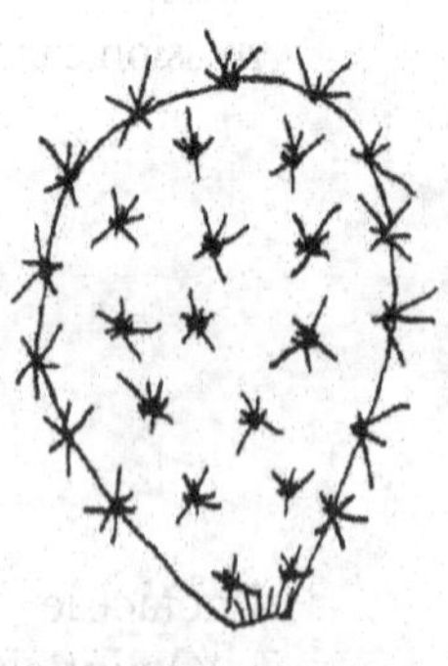

Cactaceae
Opuntia littoralis (Engelm.) Cockerell

Coast Prickly-Pear

SECTION VI
MONOCOTS

1. Perianth absent or reduced and inconspicuous. Flowers in a dense spike
   2. Leaves more than ¼ inch wide; flat. No interval between male and female flowers

      Tall perennial with thick, spongy stems 3 to 8 feet high. Leaves long linear, ⅜ to 1 inch wide, flat and sheathing at the base. Flowers numerous in a cylindrical, terminal spike 4 to 7 inches long and up to 1¼ inches thick. Spike red-brown at maturity. Common marsh plant. Blooms from June to July.

      Typhaceae
      Typha latifolia L.

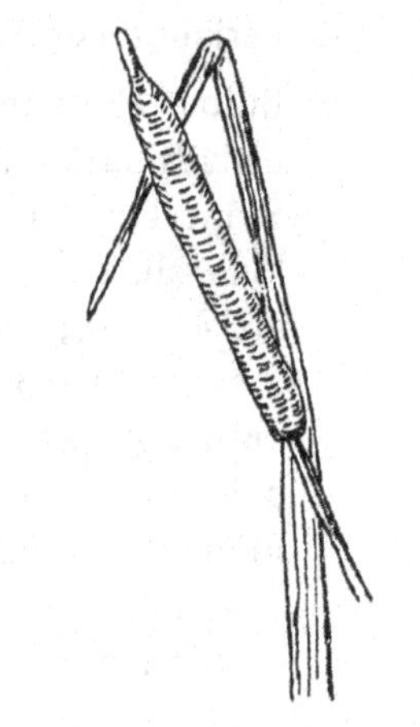

Broad-Leaved Cattail

---

   2. Leaves ¼ to ⅜ inch wide; convex. An interval up to 2½ inches between male and female flowers of spike

      Tall perennial 3 to 5 feet tall, much like the above. Leaves ¼ to ⅜ inch wide; convex on the back. Spikes 3 to 8 inches long with an interval of about 1½ to 2½ inches between male and female flowers. Common marsh plant. Often found in subalkaline water. Blooms from June to July.

      Typhaceae
      Typha angustifolia L.

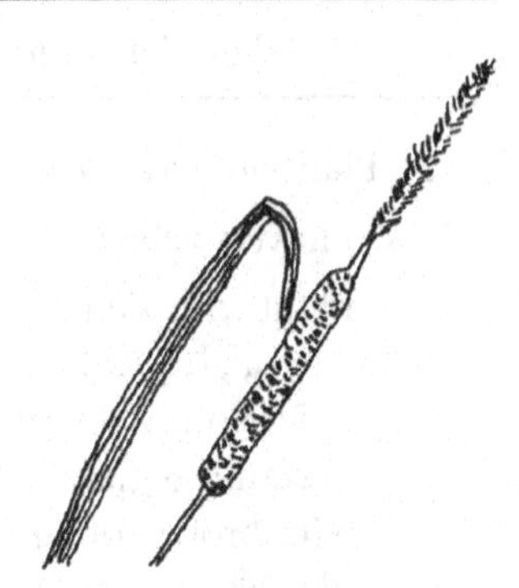

Narrow-Leaved Cattail

---

1. Perianth not absent or reduced. Segments showy and petal-like
   2. Flowers in umbels or umbellate whorls. Umbels sometimes compact and appearing globose (Flowers not in an umbel, page 194)
      3. Plant with a distinct odor and taste of onion
         4. Perianth segments red-purple; ⅜ to ½ inch long

            Plant with an ovoid bulb about ½ inch long; outer scales grayish brown and inner scales whitish. Scape 8 to 16 inches high; at the summit is an umbel of showy red-purple flowers. Leaves 2 to 4, narrow; no wider than ¼ inch. Bracts subtending the umbel somewhat scarious, ovate to lanceolate, ⅜ to ¾ inch long. Flowers 6 to 25, on ⅝ to 1¼ inch pedicels. Perianth segments 6; all alike; red-purple; about ½ inch long; acuminate with a spreading tip. Occasional on dry slopes of inner coast ranges. Blooms from March to June. **Early Onion,** *Allium praecox* Brandegee, is similar except that the ovary does not have central crests, and the bulb scales have less distinct horizontal undulations. *A. praecox* is less abundant and occurs from Ventura County southward.

            Alliaceae
            Allium peninsulare Greene

Peninsular Onion

4. Perianth segments pale rose with a darker midvein; less than ⅜ inch long

Bulb ¾ to 1¼ inches long, often in clusters. Bulb coats usually dark red; occasionally lighter or even white. Scape 4 to 16 inches high. Leaves several, 4 to 8 inches long, about 3/16 inch wide. Bracts subtending umbel are scarious; 2 to 4; joined, and thus circling the stem. Flowers 10 to 30, on pedicels ⅜ to ¾ inch long. Perianth segments 6; all alike; ¼ inch long; ovate, acute; pale rose to pink or white with a rose midvein. Common on dry chaparral slopes and moist flats. Blooms from March to May. Red-skinned onion is edible and quite flavorful.

Alliaceae
Allium haematochiton S. Watson

Red-Skinned Onion

---

3. Plant not having a distinct odor and taste of onion

4. Flowers yellow

Plant with slender, scapose stem 6 to 14 inches long and 2 basal, linear leaves ⅛ to ¼ inch wide. Flowers in a 30 to 50 flowered umbel on pedicels 1½ to 2 inches long. Perianth segments 6; all alike; yellow-orange, striped with brown along midrib. Segments spreading and star-like; about ½ inch long. Common chaparral plant but not particularly abundant. Blooms from May to June.

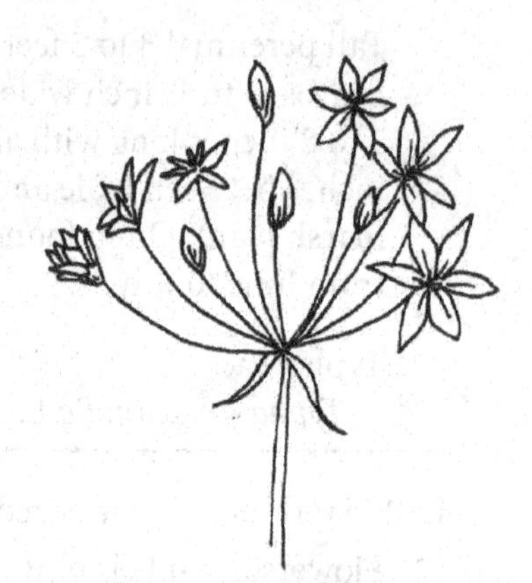

Themidaceae
Bloomeria crocea (Torr.) Coville

Common Goldenstar

---

4. Flowers not yellow

5. Flowers blue

6. Flowers in a compact umbel, head-like

Perennial herb with a slender, smooth scape 1 to 2 feet high. Leaves linear, 6 to 16 inches long and ¼ to ½ inch wide. Umbel subtended by 4 ovate, purplish bracts. Flowers in a dense, head-like cluster of 4 to 10 flowers. Pedicels up to ½ inch long, but usually less. Perianth segments 6, blue-violet, about ⅜ inch long. Anthers 6; unequal; 3 having a membranous wing extending upward as 2 lanceolate appendages that converge and conceal anthers. Common on dry hills and plains. Blooms March to May.

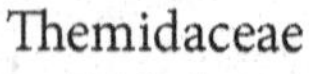

Themidaceae
Dichelostemma capitatum (Benth.) Alph. Wood

Blue Dicks

---

6. Flowers in a loose or open umbel

Stem a slender scape 2 to 10 inches high. Leaves linear, basal; longer than scape. Umbel loose or open with 3 to 11 flowers. Pedicels unequal in length, 1 to 5 inches long. Perianth segments 6, blue-violet, spreading, ⅜ to ⅝ inch long. Stamens 3, alternating with sterile stamens called staminoidia. On dry, compact, clay soil. Blooms April to June.

Themidaceae
Brodiaea jolonensis Eastw.

Mesa Brodiaea

---

5. Flowers white or greenish white

6. Flowers in regular umbels. Perianth segments 6; almost alike

Onion-like plant without odor and taste of onion. Bulb ½ to ¾ inch thick; scapes 4 to 20 inches high. Leaves narrow, linear, as long as or longer than scape. Flowers in a terminal umbel of 4 to 70 flowers. Umbel subtended by 3 to 6 scarious, lanceolate bracts. Pedicels ¼ to 2 inches long. Perianth segments 6; ½ inch or less in length; greenish white with a dark midnerve. Occasional in woods and grassland.

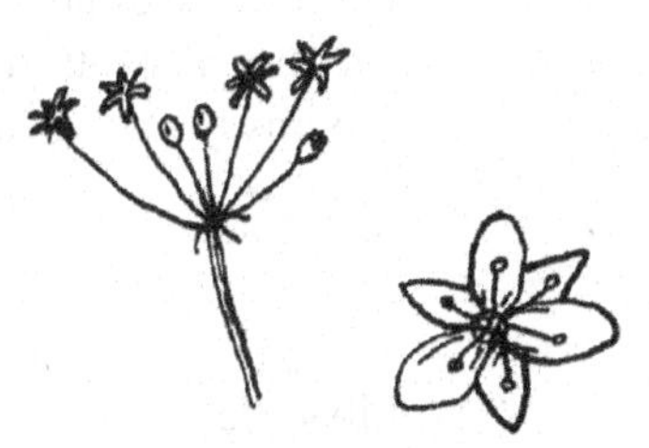

Themidaceae
Muilla maritima (Torr.) S. Watson

Common Muilla

---

6. Flowers in umbellate whorls. Perianth segments not alike. Marsh plant

Marsh plant with basal, petiolate leaves. Leaves ovate to lanceolate with a heart-shaped or truncate base. Leaf blade 1 to 6 inches long; parallel veined. Flowers in umbellate whorls on branches. Petals white, 3. Sepals green, 3. Fruit a spinose head which resembles a bur. Common marsh plant in shallow water.

Alismataceae
Echinodorus berteroi (Spreng.) Fassett

Burhead

---

2. Flowers not in umbels or umbellate whorls

3. Perianth segments 6; all alike (Perianth segments not alike, page 196)

4. Flowers blue. Grass-like plant

Grass-like, glaucous perennial 4 to 16 inches high, with a more or less winged stem. Leaves mostly basal and grass-like; up to 3⁄16 inch wide. Flowers subtended by a long leaflike bract. Perianth segments blue-violet; about ½ inch long with an emarginate apex and a pointed or mucro tip. Common plant in grassy areas throughout the chaparral. Blooms March to May.

Iridaceae

Sisyrinchium bellum S. Watson

Western Blue-Eyed-Grass

---

4. Flowers not blue

5. Flowers brown, nodding

Glabrous, slender plant 6 to 12 inches high. Leaves alternate, 2 to 5 inches long; very close together; just above ground level, narrow oblong with parallel veins. Flowers dark brown; bell-shaped; nodding. Perianth segments ¾ to 1½ inches long. Rare plant occasionally found in heavy soil or on grassy slopes. Blooms from February to June.

Liliaceae

Fritillaria biflora Lindl.

Chocolate Lily

---

5. Flowers not brown

6. Flowers orange-yellow. Leaves in whorls

Tall, stout plant 3 to 6½ feet tall. Leaves in 4 to 8 separate whorls; about 10 to 20 leaves in a whorl. Leaves oblanceolate, 3½ to 5 inches long. Flowers large, showy, nodding; few to many in a terminal raceme. Perianth segments orange-yellow with maroon spots; 3 to 3½ inches long, rolled back and under. Occasional plant of the chaparral woodland. Blooms from June to July.

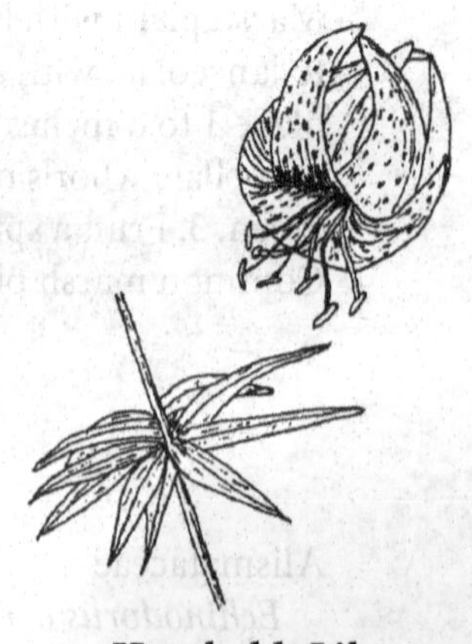

Liliaceae

Lilium humboldtii Duch.

Humboldt Lily

---

6. Flowers white to cream; sometimes with greenish veins
   7. Leaves in a basal rosette with rigid, dagger-like, pointed tips

Leaves dark green in a dense, basal rosette. Blades linear; 1 to 3 feet long, ending in a rigid, terminal spine. A rather stout, single, flowering stalk 4 to 8 feet tall rises from center of rosette. Flowers in a compact, terminal panicle 1½ to 4 feet long. Flowers creamy white, with segments 1 to 1½ inches long. Segments slightly fleshy with tips curling inward. Common on dry, chaparral slopes. Blooms April to May. Sometimes resembles lighted candles on hillsides.

Agavaceae
Hesperoyucca whipplei (Torr.) Trel.

Chaparral Yucca

---

7. Leaves not having a rigid, terminal spine
   8. Leaves mostly basal. Inflorescence a raceme

Plant with slender stem 1 to 8 feet high and basal leaves with wavy margins. Leaves reduced to bracts upward. Flowers in an elongate, terminal raceme on pedicels ¼ to ½ inch long. Perianth segments linear; ⅝ to 1 inch long, with a green midvein; spreading and curved back. Bulbs 2½ to 6 inches long; will produce a lather in water. Native Americans roasted and used them for food. The bulb may also be used as a substitute for soap. On dry hills of the chaparral. Blooms from May to August.

Agavaceae
Chlorogalum pomeridianum (DC.) Kunth

Soap Plant

---

8. Leaves not mostly basal. Inflorescence a panicle

Stout, glabrous plant with a bulb 1¼ to 2¼ inches long. Stems smooth, 1 to 3½ feet high. Lower leaves 8 to 24 inches long; ¼ to 1 inch wide; parallel veined; and folded along the midrib. Flowers in a large terminal panicle 4 to 16 inches long. Perianth segments ½ inch long; cream-colored; lanceolate ovate, with a greenish gland at the base. Occasional on dry, brushy slopes. Blooms from march to May.

Melanthiaceae
Toxicoscordion fremontii (Torr.) Rydb.

Chaparral Star Lily

---

3. Perianth segments not all alike
  4. Three outer segments sepal-like; 3 inner ones petal-like. Ovary superior
    5. Flowers yellow
      6. Gland covered with hairs. Inner surface of petals without hairs at the outer margin

Stem glabrous, 20 to 40 inches high. Leaves linear, 4 to 8 inches long, reduced above. Flowers terminal; showy. Petals bright yellow with club-shaped hairs above the basal gland and some darker markings. Gland densely covered with hairs. Petals ¾ to 2 inches long. Sepals lanceolate-ovate; 1 to 1½ inches long. On dry, chaparral hills. Common some years, scarce in other years. Blooms April to June.

Liliaceae
Calochortus clavatus S. Watson

Yellow Mariposa Lily

---

6. Gland not covered with hairs. Petals covered with long yellow hairs on inner surface. Outer margin ciliate

Perennial with stem 1 to 3 feet high and narrow, linear, basal leaves 8 to 16 inches long, withering before flowering. Flowers orange-yellow with flecks of red-brown. Inner surface of petals densely covered with long yellow hairs. In chaparral from Santa Ana Mountains to San Diego. Flowers from May to July.

Liliaceae
Calochortus weedii Alph. Wood

Weed's Mariposa Lily

---

5. Flowers white to pale lavender or dark purple
  6. Flowers all white, globose, nodding

Plant with slender, erect stem 8 to 32 inches high. Basal leaf 1 to 2 feet long and ½ to 2 inches wide. Upper leaves linear lanceolate, 2 to 10 inches long. Flowers pure white, globose, and nodding. Petals ¾ to 1 inch long with ciliate margins and white hairs above gland. Delicate and beautiful; in shaded areas of chaparral. Blooms from April to June.

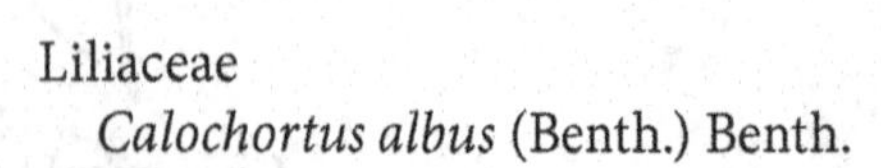

Liliaceae
Calochortus albus (Benth.) Benth.

White Globe Lily

---

6. Flowers lilac, pinkish-rose, or white tinged with lilac; bowl-shaped, erect
   7. Petals pinkish rose; inner surface covered with long, yellow hairs

      Perennial with stems 1 to 2 feet high and basal leaves narrow linear, 8 to 16 inches long and about ½ inch wide, withering at flowering time. Leaves on main stem reduced. Flowers bowl-shaped with pinkish-rose petals which are covered with long, yellow hairs. Gland is circular, slightly depressed and surrounded with hairs. Found on dry rocky slopes and brushy areas below 5,000 feet elevation. Santa Monica Mountains to near San Jacinto Mountains. Blooms from May to July. **RARE.**

Liliaceae
Calochortus plummerae Greene

Plummer's Mariposa Lily

---

7. Petals not pinkish-rose and covered with long yellow hairs
   8. Gland, at base of petals, covered with thickened fungus-like hairs

      Perennial with stem 8 to 24 inches high. Basal leaves 4 to 6 inches long and about ¼ inch wide. Leaves on main stem reduced. Flowers deep lilac, often with a purple spot. Gland covered with branched, fungus-like hairs. Occurs in coast ranges and on dry slopes in heavy soil. Not common. Blooms May to June.

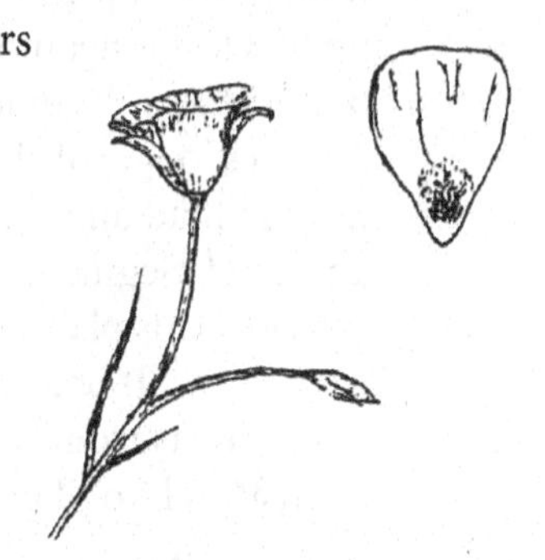

Liliaceae
Calochortus splendens Benth.

Splendid Mariposa Lily

---

8. Gland covered with thin hairs
   9. Petals white, tinged with lilac; maroon colored spot at base of petal around the gland. Santa Barbara to San Diego Counties

      Erect, slender plant 8 to 24 inches high. Leaves linear, much reduced upward. Basal leaves withered at flowering time. Flowers showy and terminal; bowl-shaped; white; lilac tinted near tips. Petals broadly ovate, rounded or obtuse above; maroon spot above gland at base; gland densely covered with hairs. Sepals lanceolate, often with a maroon spot at base. Common spring plant in dry hills of the chaparral. Blooms March to May.

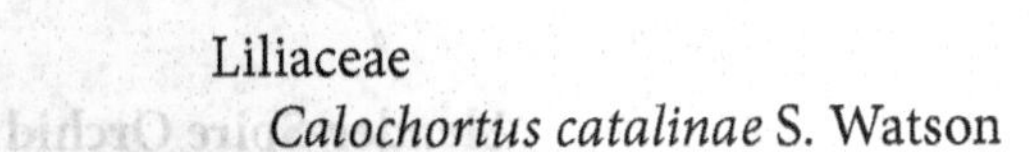

Liliaceae
Calochortus catalinae S. Watson

Catalina Mariposa Lily

---

9. Petals white to lilac or red-purple with a dark red blotch in middle. Los Angeles County and north

Perennial with linear, basal leaves 4 to 8 inches long and less than ¼ inch wide. Leaves on main stem reduced. Flowers usually white to dark red with a dark red blotch in middle and sometimes another above it. Gland covered with hairs. Inflorescence forming an umbel but not obvious since there are only 1 to 3 flowers. Sepals lanceolate; usually curled back at tip. In coast ranges and in sandy soil; northern Los Angeles County northward. Flowers May to July.

Liliaceae
Calochortus venustus Benth.

Butterfly Mariposa Lily

---

4. Flowers not as above. Ovary inferior
5. Flowers large, showy; blue to purple

Main stalk of plant 6 to 32 inches high. Leaves linear; up to 3 feet long; mostly basal. Flowers subtended by 2 opposite leaflike bracts 2 to 4 inches long. Petals 1½ to 3 inches long; pale blue to purplish. Three outer segments broad and spreading; the 3 inner ones narrower and erect. Segments united below into a tube. Ovary inferior. This plant is not found in the Santa Monica Mountains. It occurs in the mountains and on coastal grassy areas from Santa Barbara northward. Blooms from March to May.

Iridaceae
Iris douglasiana Herb.

Douglas' Iris

---

5. Flowers small, irregular, not blue to purple.
6. Flowers with a basal spur. Plant of chaparral woodland.

Erect, glabrous plant 1 to 3½ feet high. Leaves 2 to 4; linear lanceolate; 2½ to 8 inches long. Flowers small, green, in a terminal spike 4 to 12 inches long. Sepals 3. Petals 3; 2 alike, and the third forming a spurred, basal sac. Relatively rare plant of the chaparral woodland. Blooms from April to August.

Orchidaceae
Piperia unalascensis (Spreng.) Rydb.

Slender-Spire Orchid

---

6. Flowers without a basal spur. Plant of moist stream bank

   Erect perennial herb, 1 to 3 feet high. Leaves lanceolate, 2 to 6 inches long, reduced upward. Flowers small, in a raceme, subtended by leaflike bracts. Flowers irregular with 3 sepals and 3 petals. Sepals green, and petals greenish to yellow, tinged with red. Stream orchid is occasionally found on moist stream banks. Blooms from March to June.

   Orchidaceae
   Epipactis gigantea Hook.

Stream Orchid

INTRODUCTION TO THE FLOWERING PLANTS

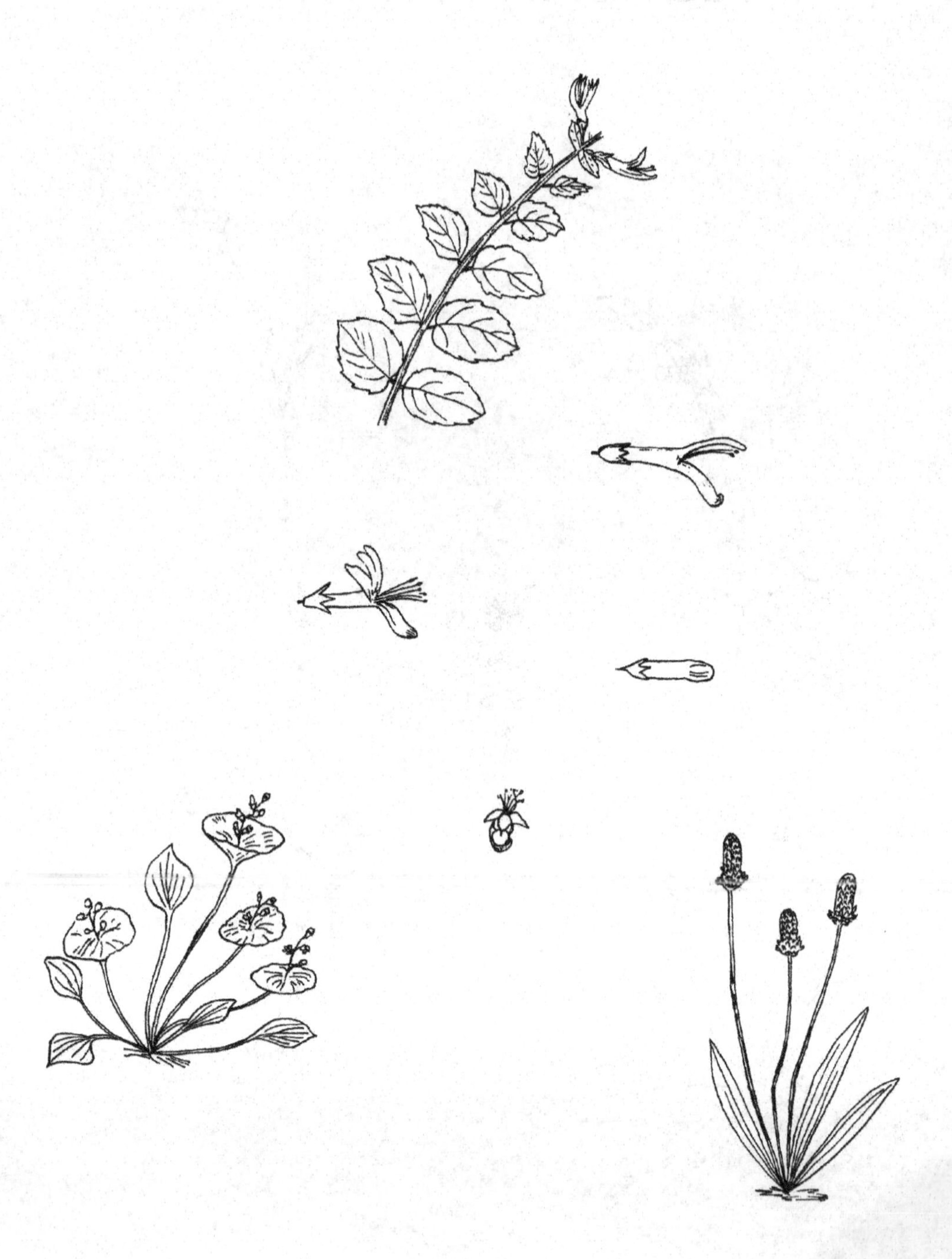

ANGIOSPERMS

Flowering plants, known as **angiosperms**, have their reproductive structures in flowers. In these structures seeds are produced which are enclosed in fruits. On the basis of seed types, angiosperms are divided into two main groups: **monocotyledons** and **dicotyledons**. Monocotyledons, or monocots, are plants which have one cotyledon or **seed leaf**. Dicotyledons, or dicots, have two cotyledons or seed leaves. Fortunately, other differences occur between the two major groups, besides their seed types, so that dissecting a seed is not necessary to determine if a plant is a monocot or a dicot. Monocots are plants that have parallel-veined leaves, whereas dicots have net-veined leaves. Flowers of monocots have three to six petals, but never four or five. In comparison, most dicots have flowers with four, five, or more petals, but rarely six. Dicots that have flowers with six petals have leaves that are not parallel-veined. Monocots include the grasses, lilies, iris, and orchids, as well as many other plant species.

Monocots	Dicots
3 to 6 petals	4, 5, or more petals
Parallel-veined leaves	Net-veined leaves
One cotyledon	Two cotyledons

Stems

Stems of plants are quite variable and may be useful for purposes of keying out plants. Section I is separated on the basis of the stem being woody; trees and shrubs. Section II includes herbaceous plants (not woody). Herbaceous plants may be annuals (live only one year), biennials (live for two years), or perennials (live for many years). Perennials are usually larger than annuals but are best identified by the presence of some woody tissue at the stem or in the upper root.

The position of the stem with respect to the ground is also variable. Some stems are **prostrate** (lie flat on the ground), while others are **ascending** (grow upward at a angle to the ground) or are **erect**.

Leaves

The leaf is a very important part of the plant and is often used as a keying characteristic. Between the leaf and the stem is an **axillary bud**. The expanded portion of the leaf is known as the **blade**, and the stem base is known as a **petiole**. Occasionally leaves have no petiole and are known as **sessile** leaves.

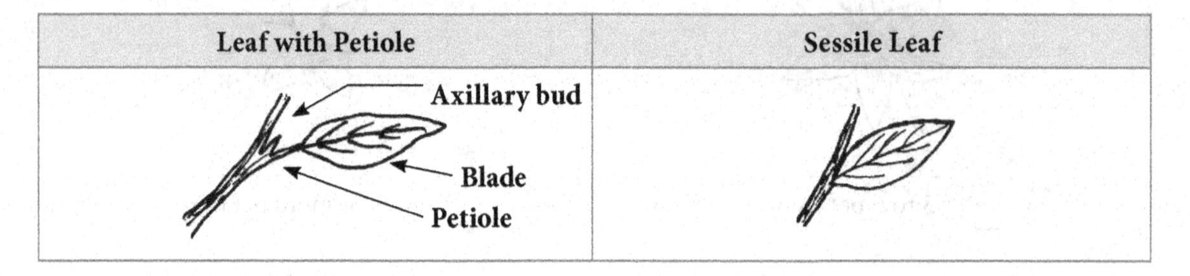

Some species of plants have paired glands, scales, or leafy structures at the base of the petiole. These are known as **stipules**.

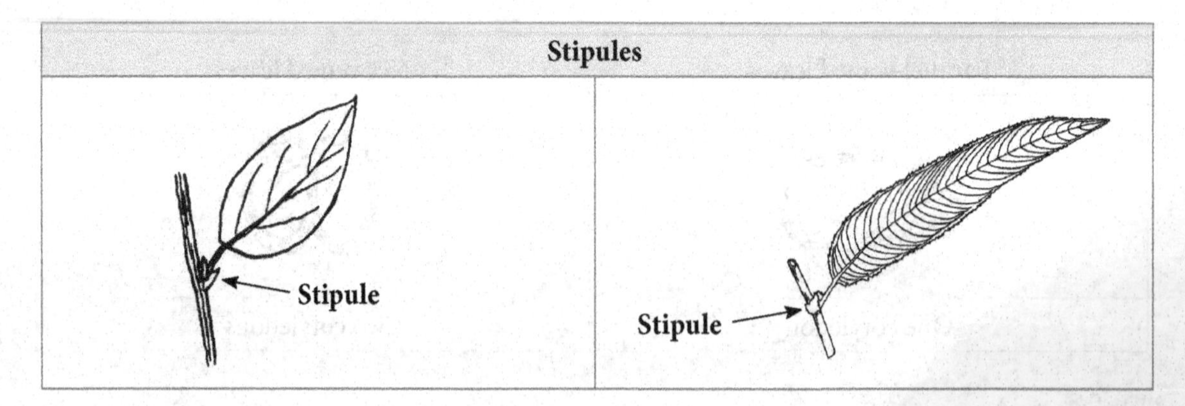

The arrangement of leaves on the stem is also important. Leaves may be **opposite, alternate**, or **whorled**. Whorled leaves have more than two leaves arising from the same point or **node** on the stem. **Basal leaves** occur only at the base of the stem. Extending above the basal leaves may be a naked flowering stem called a **scape**.

Opposite Leaves	Alternate Leaves	Whorled Leaves	Basal Leaves/Scape
		 Node	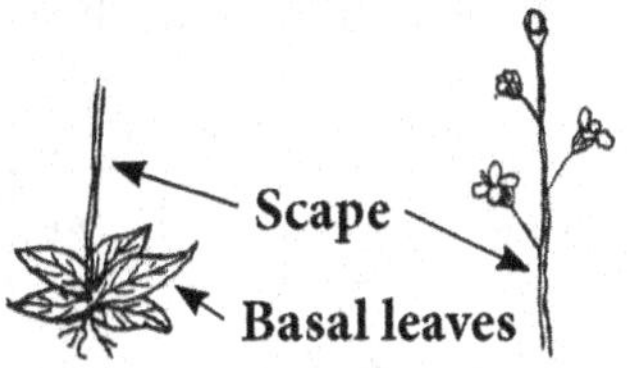 Scape Basal leaves

Venation of leaves can also be used for keying purposes. **Parallel-veined** leaves are typical of the monocots. Dicots have **net-veined** leaves. Net-veined leaves may be **pinnately veined** or **palmately veined**.

Parallel Veined	Pinnate Veined	Palmate Veined

Shapes of leaves vary considerably. Some of the common shapes are shown below. Sometimes a leaf is neither ovate nor lanceolate, but seems to lie somewhat in between. The shape is then indicated as ovate-lanceolate. Occasionally, the prefix **ob-** might be used in front of the shape name in such case.

Leaf Shapes				

Filiform	Obovate	Spatulate	Oblong	Orbicular

Ovate	Lanceolate	Linear	Deltoid	Heart-shaped

Leaf margins vary with teeth (**toothed** or **dentate**), **lobed**, or **entire** (without teeth or lobes). There is variation with the type of teeth at the leaf margin. A leaf may be coarsely or finely toothed. Sometimes the teeth are rounded, with a **crenate** margin.

Lobing of a leaf may also vary. A leaf may be **shallow lobed** or **deeply lobed**. The leaf is said to be **parted** if it is deeply lobed, and **divided** if it is lobed almost to the main axis. A few of the common margin types are illustrated below.

Leaf Margins		
		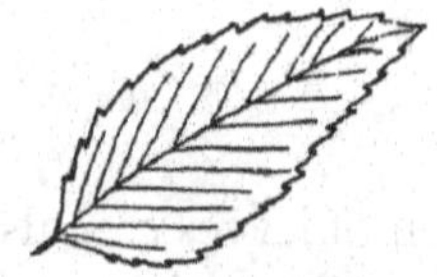
Undulate	Coarsely toothed	Finely toothed

Parted	Palmately lobed	Crenate

Leaves may be simple or compound. **Simple leaves** have only one blade. **Compound leaves** have more than one blade, and these blades are called leaflets. To determine if a leaf is simple or compound, it is best first to locate the **axillary bud**. If there is just a single blade arising from beneath the axillary bud, it is a simple leaf. If more than one blade is on the leaf arising from beneath the axillary bud, the leaf is compound. All **leaflets** of a compound leaf lie in one plane and will lie flat if you will place them on a flat surface, such as a table. Simple leaves will not lie flat because they are spirally arranged on the plant.

Simple Leaf	**Compound Leaf**
 **Axillary bud**	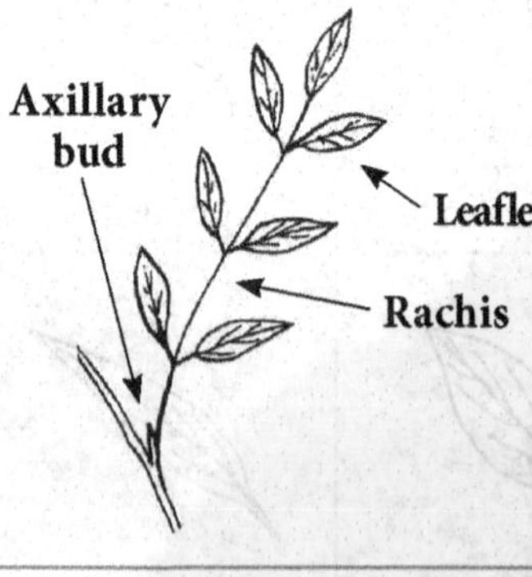 **Axillary bud** **Leaflet** **Rachis**

Compound leaves may be pinnate compound or palmate compound. **Pinnate compound** leaves have leaflets on either side of a main axis, somewhat like a feather, while **palmate compound** leaves have the leaflets all radiating out from a point. Pinnate compound leaves can also be two times pinnate or twice compound. Some leaves are even three times pinnate.

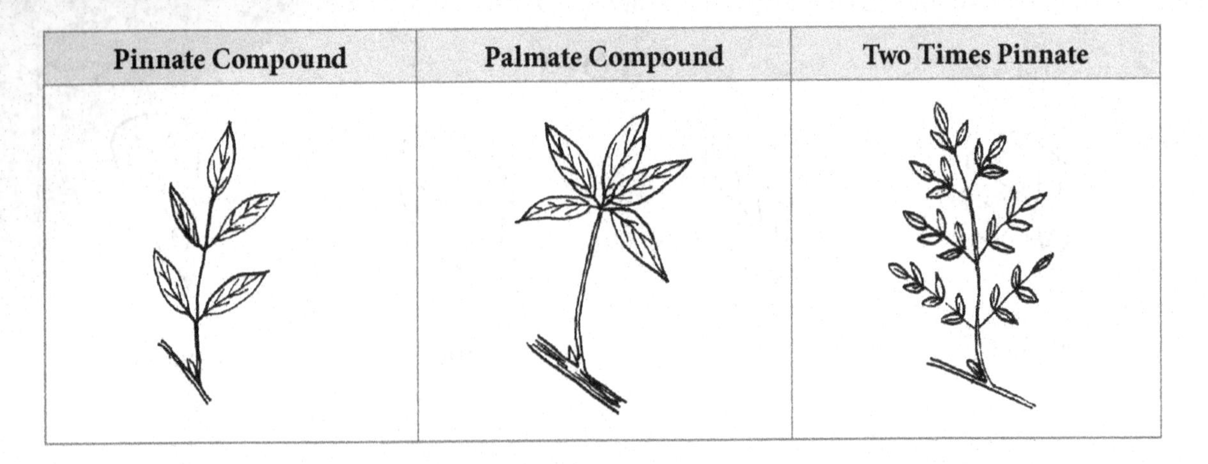

When a compound leaf has three leaflets, the term **trifoliate** or **3-foliate** is often used. A good example of a trifoliate leaf is the common clover.

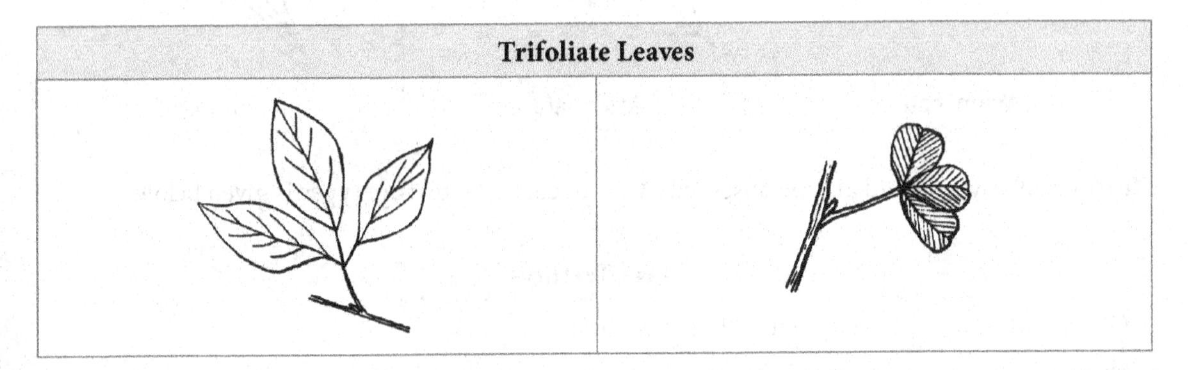

In the key, the term **pinnatifid** is frequently used. A pinnatifid leaf is a simple leaf that is deeply, pinnately divided (lobed) almost to the main axis. It is almost a compound leaf, but not quite. A pinnatifid leaf may also be once-, twice- or three-times pinnatifid.

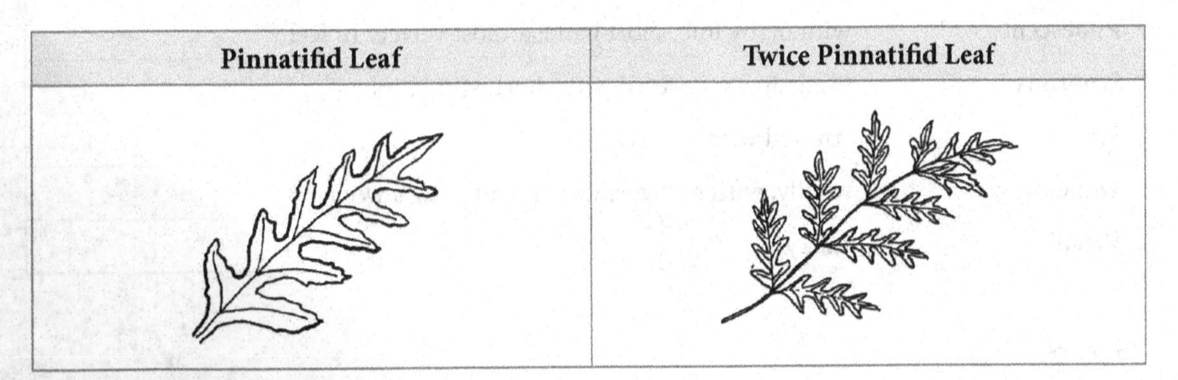

The **leaf apex** is also of importance in keying. The apex may be **rounded, acute, obtuse,** or **truncate** (chopped-off appearance). An apex more pointed than acute is **acuminate** or **attenuate**. Occasionally, the leaf apex is **spinose,** as in yucca.

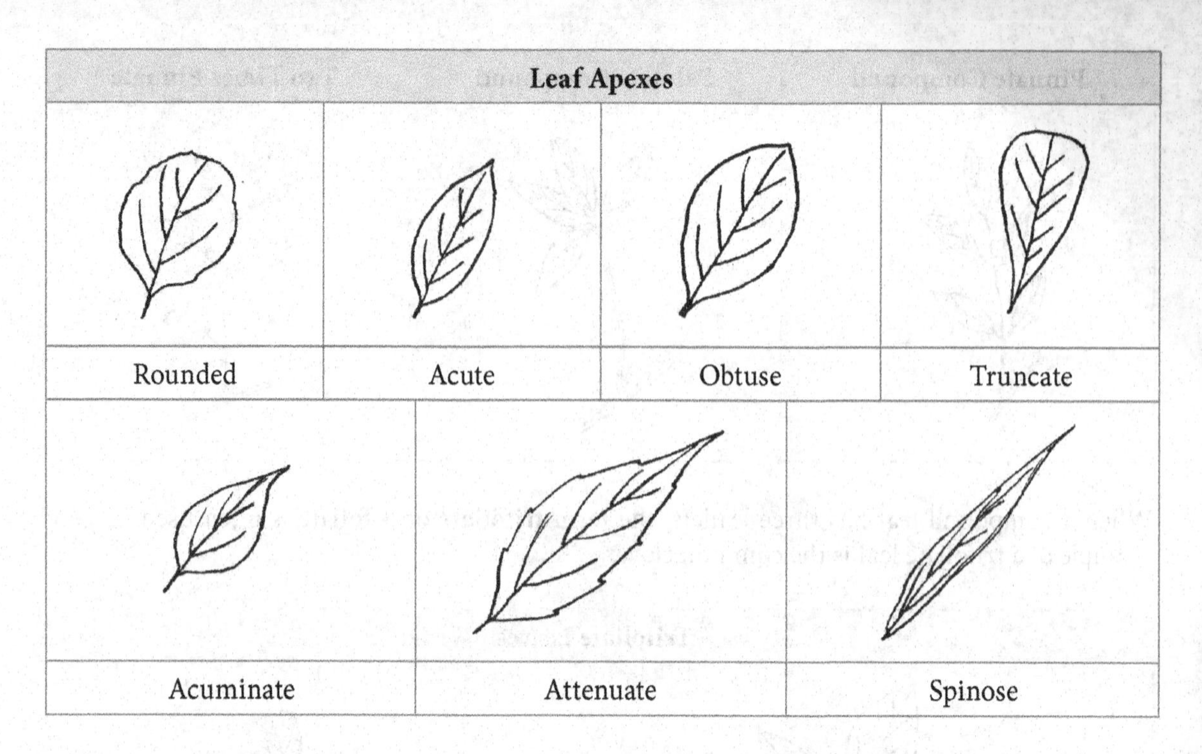

Textures of leaves should also be observed. A list of common texture types is given below.

Leaf Textures	
Canescent	covered with grayish hairs
Glabrous	without hairs
Glandular	with tiny glands which contain a sticky substance
Hispid	with stiff or bristly hairs
Pubescent	with many fine, short hairs; almost velvety in feel
Scabrous	sandpapery; covered with short, stiff hairs
Scurfy	covered with tiny scales
Tomentose	woolly; with many wavy hairs forming a wool
Viscid	sticky

Flowers

The **flower** is the reproductive structure of the plant. Some flowers are large and showy and have bright colored petals whereas others are very small, greenish, and inconspicuous. Brightly colored flowers attract insects which carry pollen from flower to flower in a process called **pollination**. Small inconspicuous flowers are pollinated by wind.

The essential parts of a flower are the male and female structures. The **pistil**, consisting of **stigma**, **style**, and **ovary**, is the female portion. The stigma is the sugary, upper portion to which the pollen sticks. The ovary is the slightly swollen base where fertilization takes place. Inside the ovary are one or more tiny seed-like structures called **ovules**. Each ovule contains one **egg**, and if fertilized, the ovule develops into a **seed**. The ovary may swell many times its original size and becomes the fruit in which the seeds are enclosed.

Stigma
Style
Ovule
Ovary
Pistil

The male part of flower is the **stamen** which consists of a **filament** and an **anther**. **Pollen** is formed in the anther, and during pollination is picked up by wind or insects and carried to the stigma of the pistil. After pollination, the pollen begins to germinate. Stimulated by the sugary substance on the stigma, one cell of the pollen grain grows a small tube which passes down the style of the pistil and ultimately contacts an ovule in the ovary. More than one pollen grain can grow down the style at the same time. While the pollen tube is growing, the other cell of the pollen grain divides to produce two **sperm**. It is one of these sperm that contacts the **egg** in the **ovule**. When the egg and sperm fuse together, **fertilization** has been accomplished, and a new plant has been produced. The fertilized egg grows into an embryo, all of which is still retained within the ovule. The inner tissue of the ovule becomes **endosperm** or nutrient tissue for the embryo, and the outer tissue hardens to form a protective **seed coat**. A mature seed consists of a partly grown embryo (baby plant) embedded in a nutrient tissue, all of which is surrounded by a resistant coat. The seed is a marvelous adaptation and is able to survive all types of adverse environmental conditions. Most seeds live from 7 to 25 years, but many have been known to live over 100 years, and some for more than 1,000 years.

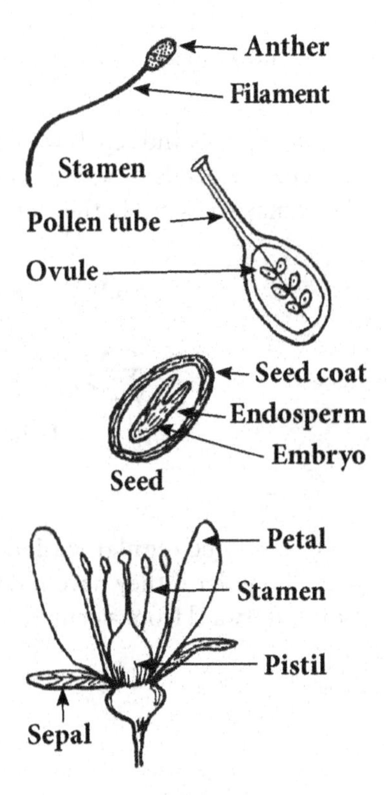

Although the pistil and stamens are the only essential parts of a flower, most flowers also have **sepals** and **petals**. In most plants, the petals are larger and more brightly colored than the sepals. Sepals are generally green. In some plants the flowers have sepals but no petals. When petals are absent, the sepals may be colored and almost look like petals.

The shape, number, and color of petals is extremely variable. Petals can be fused into one structure, called a corolla. A corolla can be shaped like a funnel (**funnelform**) or be **bell-shaped, urn-shaped, salverform** (like a cake plate) or **tubular**.

Corollas				
Funnelform	Bell-shaped	Urn-shaped	Salverform	Tubular

Sepals may also be fused into one structure, called a **calyx**. The calyx may also be a variety of shapes, such as: cup-shaped, cylindric, tubular, or two-lipped.

Calyxes			
Cup-shaped	Cylindric	Tubular	Two-lipped

Fusion of petals and sepals is rarely complete. In many plants, **teeth** or **lobes** occur which represent the separate petals or sepals which have not fused. The number of these teeth or lobes for many plants may be useful in their identification.

Corolla Lobe	Calyx Teeth
Lobe	Teeth

Flowers may be **regular** (**radial symmetry**) or **irregular** (**bilateral symmetry**). Flowers are considered regular if they have radial symmetry. In regular flowers, the petals are the same size and radiate outward from a central point. Buttercup, rose, and wild mustard are regular flowers.

Regular Flowers (Radial Symmetry)		
Buttercup	Rose	Wild mustard

Irregular flowers have bilateral rather than radial symmetry, and petals which are not equal in size. The irregular flower, if divided down the center, is laterally symmetrical so that the right side is a mirror image of the left side. Snapdragon, sweet pea, and mint are examples of irregular flowers.

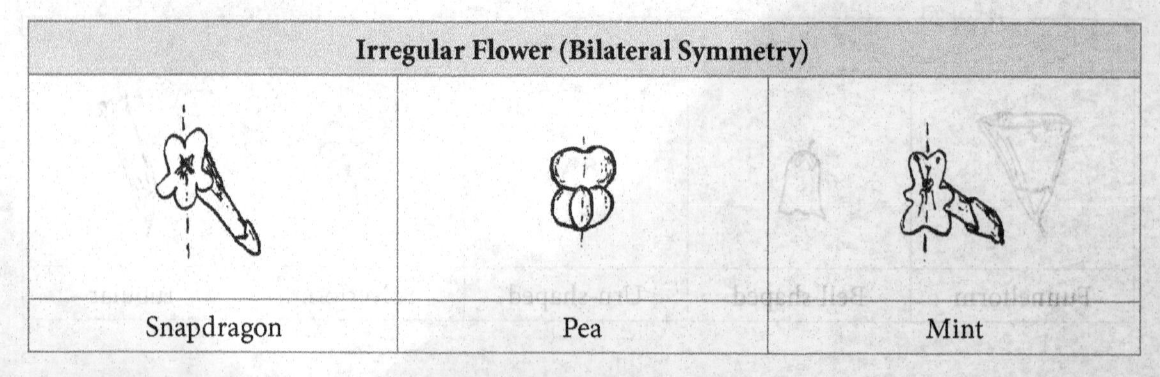

Irregular Flower (Bilateral Symmetry)		
Snapdragon	Pea	Mint

A rather common type of irregular flower is one which is **two-lipped**. Two-lipped flowers are those with an upper lobe and a lower lobe. Snapdragon, monkey flower, and the mints are all two-lipped flowers.

Two-lipped Flowers
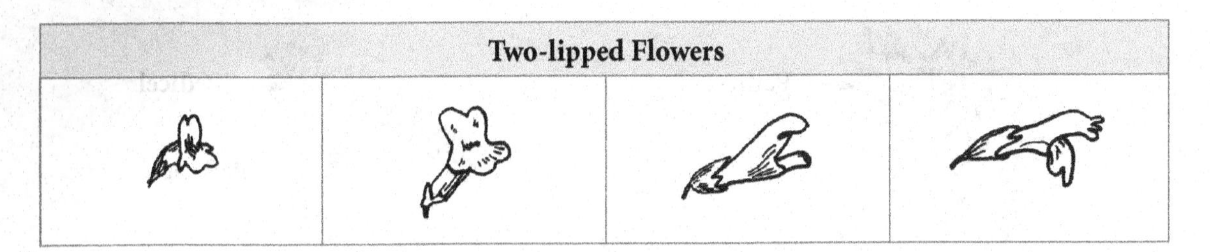

The pea flower is another type of irregular flower. It is unique and occurs only in the pea family. It is called a **papilionaceous flower**.

Papilionaceous flower

In different kinds of flowers, the position of the ovary relative to the petals may be different. In those flowers in which the ovary is positioned above the petals and sepals, it is called a **superior** ovary, and this is the most common type. In some flowers, the ovary is surrounded by the petals and sepals and appears to be about in the middle. This ovary position is called **perigynous** and is found in many members of the rose family. The third type of ovary is **inferior** in which the ovary is beneath the petals.

Types of Ovaries		
Superior ovary	Perigynous ovary	Inferior ovary

The way in which the flower is placed on the stem also may vary in different plant species. A flower may be **solitary** or occur in a flower cluster called an **inflorescence**. Solitary flowers are either **terminal** (located at the tip or summit of the stem) or **axillary** (arising from the leaf axil).

Flower Position	
Terminal	Axillary

The stem that supports the flower is known as a **pedicel** and may be long or short. A stem that supports a flower cluster (inflorescence) is called a **peduncle**.

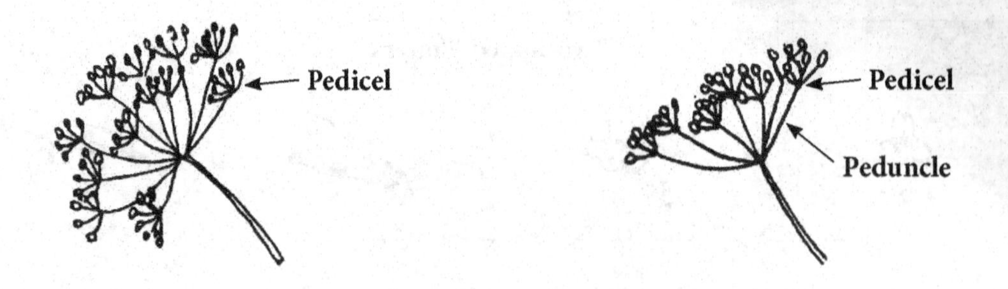

Inflorescences may also be axillary or terminal. They can be many-flowered or relatively few-flowered, dense and compact, or loose and open. Some of the more common types of inflorescences are shown below.

Inflorescences				
Raceme	Spike	Head	Umbel	Panicle

A **catkin** is a unique type of inflorescence, consisting of a spike of unisexual flowers. Plants with catkins can include cottonwood, willow, oak, alders, and walnut trees. Most catkins are nothing more than spikes of stamens or spikes of ovaries.

Male Catkin	Female Catkin

At the base of the inflorescence there may be one or more leaf-like structures called **bracts**. A cluster or whorl of bracts is known as an **involucre**.

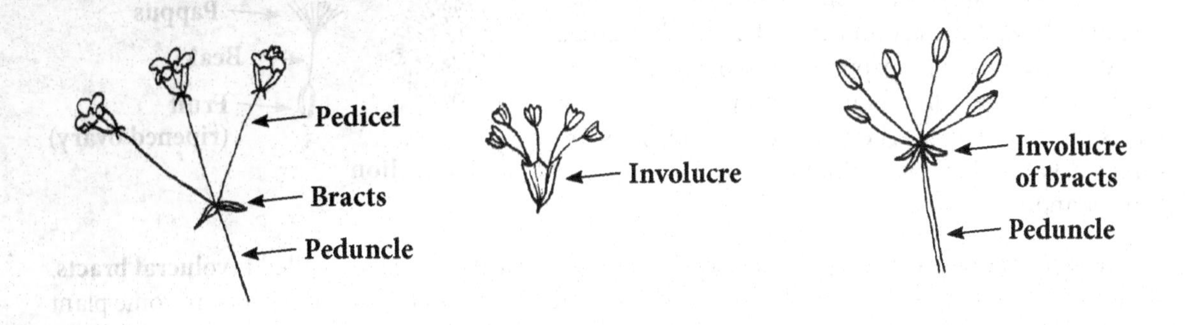

Composite Flowers

Members of the **Asteraceae** family, also known as composite flowers, such as daisy and dandelion, are a special group of flowers in which the flower is not really one flower but many. The flowers are compacted into a dense inflorescence called a **head**. The types of flowers in the composite head are also unique. Two types occur: **ray** and **disk** flowers. **Ray flowers** consist of one broad petal and an inferior ovary. Stamens may or may not be present. Ray flowers form the outer petal-like structures of the daisy. A dandelion consists of all ray flowers.

Sunflower	Dandelion	Ray Flower
		Corolla Pappus Ovary

Disk flowers are the other type of composite flower, and they form the center part of the daisy. Most disk flowers are perfect, having both stamens and pistil. The ovary is inferior, just as in the ray flower. Composites may have both ray and disk flowers, only ray flowers, or only disk flowers. A thistle, for example, has flowers consisting of only disk flowers.

Thistle	Disk Flower
	Stigma and Style Stamen Corolla Pappus Ovary

In some composite flowers, the ray flower has a structure just above the ovary and at the base of the corolla called as **pappus**. The pappus may consist of many capillary or thin white hairs, or may be a series of scales or thin bristles. The pappus of dandelions are capillary hairs which most children have blown at one time or another.

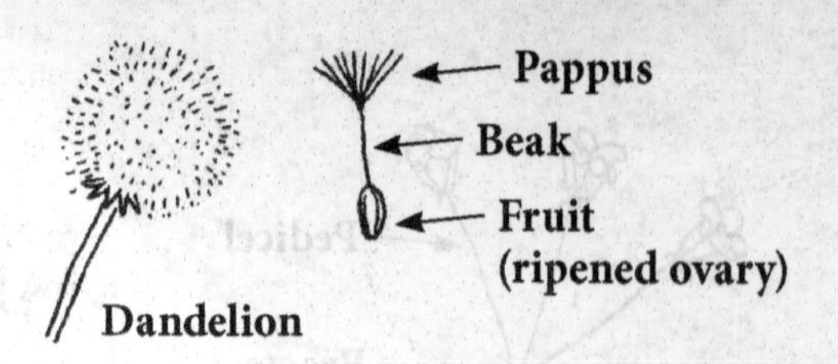

On the outside of the composite head are a series of green leafy structures called **involucral bracts**. In most of the composite plants, the bracts overlap one another in two or more rows. In some plant species, the bracts are not green, but are white and papery. Pearly everlasting is an example of this. The "everlasting" parts of the "flowers" are really the involucral bracts which open out and look like papery petals.

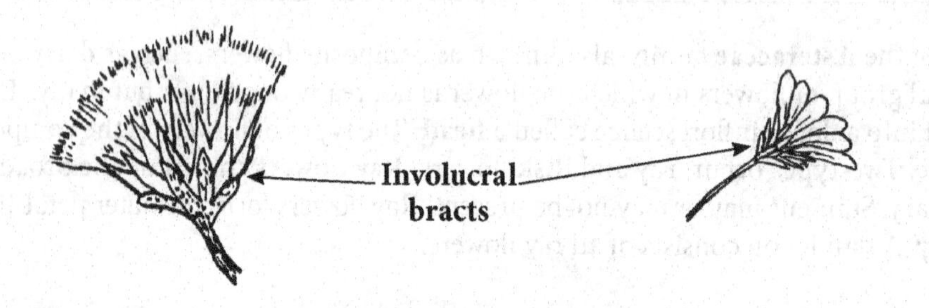

Fruits

The fruit is the ripened ovary of the flower. After fertilization, the ovary enlarges several times its original size. Inside the fruit are the seeds.

Fruits are as varied as flowers and may be dry or fleshy and juicy. A **nut** is a dry fruit and a **berry** is a fleshy fruit. Dry fruits may open at maturity and release their seeds or may remain closed and be dispersed with the seeds. Fruits that do not open are usually one-seeded and commonly small. Sunflower "seeds" (really fruits) and grain (wheat, oats, barley, rye, etc.) are dry fruits with one seed. Grain is a fruit in which the seed is actually fused to the fruit.

Dry fruits that open at maturity include the **follicle, legume**, and **capsule**. The follicle opens on one side (milkweed fruit); the legume opens on two sides (lima bean, pea, green bean, peanut); and the capsule opens along three or more sides (cotton).

Dry Fruits		
Follicle	Legume	Capsule

If the pea pod is the fruit (a legume in this case), the peas which we eat as vegetables are really seeds inside the fruit. Slice a pea lengthwise sometime and observe the tiny embryo inside. Corn is a grain in which the fruit and seed are fused. The embryo on the corn kernel is the little white part at the base.

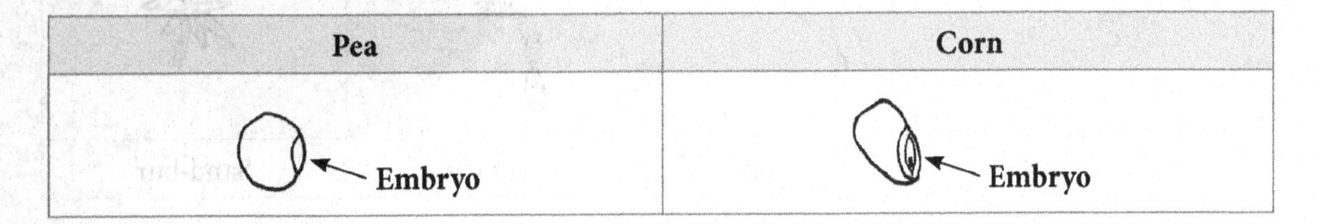

Pea	Corn
Embryo	Embryo

Fleshy fruits include the berry, drupe, and pome. The berry is entirely fleshy, like the huckleberry and tomato in which the seeds occur inside the flesh. The drupe differs from them as it has a stony, inner portion or shell which encloses the seed (peach, cherry, plum). The pome is peculiar as in that the core portion is the ripened ovary, whereas the fleshy, edible part develops from the base of the flower (apple). Remnants of the floral parts (stamens and sepals) can be seen at the end of the apple opposite from the stem.

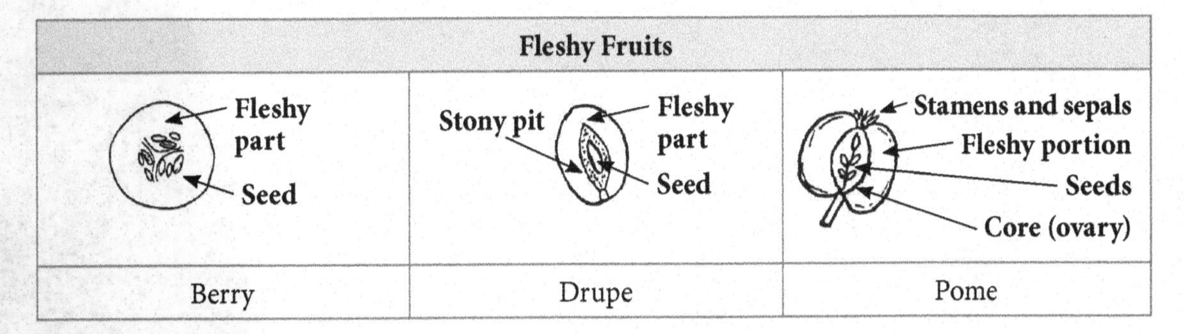

Fleshy Fruits		
Fleshy part Seed	Stony pit Fleshy part Seed	Stamens and sepals Fleshy portion Seeds Core (ovary)
Berry	Drupe	Pome

Raspberry and blackberry are not true berries, even though the name implies this. Actually they result from a fusion of many ovaries rather than just one and thus form an **aggregate** fruit. Each rounded bump on the blackberry or raspberry represents a single, matured ovary.

Seeds

Seeds are located inside the fruit and represent the ripened ovule. Inside the seed is the partially grown embryo of the new plant. The embryo is embedded within nutrient tissue so that the young plant will have enough food to survive the initial stages of growth before it reaches the surface of the soil and starts to produce its own food.

Seeds are either released from the fruits at maturity or are dispersed or disseminated along with the fruit. Plants have many fantastic adaptations to enable fruit and seed dispersal. Wind is the most common agent, but insects and animals are also important.

Fleshy fruits are attractive to animals who then disperse the seeds. Some fruits have burs or spines to catch on clothing or fur and are thus carried many miles. Winged fruits or seeds are quite common as also are seeds with cottony tufts of hairs. These seeds may be carried many miles in the wind. Seeds of cottonwood, cotton, milkweed, and dandelion are all examples of this type of seed dispersal.

Seed Dispersal			
Cocklebur	Dandelion	*Clematis*	Sand-bur

After seeds are dispersed, they may become covered by soil and lie dormant there until the environment becomes favorable for germination. Many seeds, particularly those of the northern United States and Canada, for example, will not germinate unless the seeds have first been through one or more periods of freezing and thawing. This prevents the seeds from sprouting in early October during a warm "Indian summer."

HOW TO COLLECT, PRESERVE, AND MOUNT PLANTS

Collecting

You may be interested sometime in collecting and mounting pressed plants that you have identified. Even weeds in your yard make nice specimens when pressed and mounted. When collecting plants, be sure to obey the laws concerning picking. Plant collecting is prohibited in most state and national parks. Collect plants only in regions where it is permissible and do not pick more than you intend to press and use. Select an average specimen. Pick only what is abundant and not protected by state laws.

The first thing to consider is how much of the plant you are going to collect. If the plant is large, take a representative sample, including flower and leaf. If the plant is small, you may wish to press it all. When you collect a small plant for scientific purposes, the entire plant is necessary, including the roots.

After you have collected your specimens, they must be pressed as soon as possible to prevent them from shriveling or curling up. Some plants maintain their form better than others. If the plant is from a dry area, the picked specimen will probably last longer than one from a shaded stream area. If you will not be able to press your plants immediately, you should bring along a plastic bag in which to keep the plants. If placed in the bag with a moist paper towel, the picked plants will remain fresh for quite a while. If, upon arriving home, you still cannot press the plants at once, place them in the refrigerator until you can get to them. But remember, the sooner they are pressed, the nicer they will look.

Pressing

You will probably want to construct some sort of plant press if you are going to do much pressing of plants. A very satisfactory plant press can be made by using two flat boards or by hammering strips of boards together. The size of a professional plant press is 12 inches by 18 inches. If you make one of this size, pressing the plants within folded newspapers is very easy. You will also need one or two straps to clamp the press together.

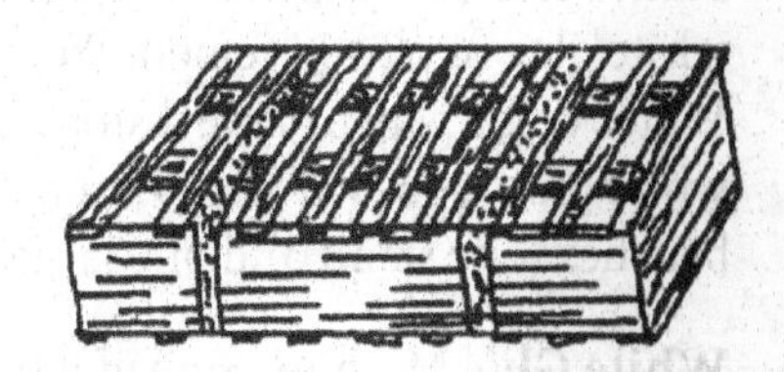

Plant press

Plant presses may also be purchased from any biological supply house. The prices range depending upon the size and quality.

In order to facilitate drying, you should separate the specimens between corrugated cardboard (ventilators). These may be obtained easily by cutting cardboard cartons to the correct dimensions. The professional presses also come equipped with blotters (dryers). Newspapers, however, may be used in place of blotters. Because newspapers are less absorbent than blotters, the newspapers will have to be changed once or twice each day to speed up the drying process and to prevent the plants from becoming moldy.

When you have the press and plants ready, carefully place the plant between a single folded sheet of newspaper. When placing your plant, try to flatten out the leaves and flower petals. The way you place your plant is the way it will remain after it is pressed. When the plant is correctly placed, lay the top fold of the newspaper over it carefully, pressing down with the hands. Next, put one of the corrugated cardboards (ventilators) on top, continuing to press down with the hands. Place a dryer on top of the ventilator and then proceed with the pressing of another plant specimen. Continue to alternate plants between the ventilators and dryers. When all plants have been pressed, place the stack between the two boards and pull together firmly, using the straps.

The plants must now be dried. There are several ways in which they may be dried, depending upon the time of the year and the place in which you live. If it is summer time and the sun is hot, you may place your press out in the sun. Be sure to bring it inside at night or if it starts to rain. If the weather is colder, but still sunny, you may place your press in a closed car which will heat up sufficiently to dry the plants.

If you wish to dry your plants indoors, you may use a desk lamp for heat. Do not use too strong a bulb or get the lamp too close to the press as you will burn the plants, causing them to discolor and turn brown. If you have a hair dryer, that may be used too. Air and heat are desirable, but the press should only be warm, **never hot**. Most plants will dry in three or four days. If the plant is unusually succulent, drying will take much longer. Cactus may take several weeks to dry completely. To speed the drying process of succulents, slice the plant longitudinally and dig out some of the succulent, center portion with a knife. Laying the plant in a dish of salt overnight will also help to draw out some of the water.

When your plants are thoroughly dried, they should be removed from the press. Be sure that they are sufficiently dried. If plants are not completely dry, they will curl up and not remain flat. If left to dry for too long, or if they are dried at too high a temperature, they will darken, and the flowers will lose their color.

Mounting

Plants can be mounted in several ways. Two of the most popular mounting materials in use are white glue or plastic cement. You will also need some white mounting paper of the appropriate size. Water color paper or card stock is good to use and comes in a variety of sizes. If you are using the white glue, you will also want a glass plate (12 inches by 12 inches or larger), and a couple of paint brushes (one small and one large). An oil gun is useful if you plan to use the plastic cement.

White Glue Method: Spread a thin layer of glue on a glass plate. Thin the glue with water to make it spread easily. Next, lay the pressed plant on the glass, the wrong side down, so that the plant will pick up the glue. Slide the plant around slightly, then pick it up and place it on the sheet of mounting paper. Decide before hand exactly where you wish to place the plant, because once the plant is placed, the glue on the paper will prevent you from moving the plant around. If you have some tweezers, the plant can be picked up more easily from the glass plate than by hand. To hold the plant down on the paper, use some sort of weights. Large metal washers are extremely useful because the holes can be placed over parts that have glue to which the washers might otherwise stick. If there are any small leaves or petals which do not stick to the paper, a small paint brush can be used to "paint" some glue on the underside before fastening down. When the glue is dry, remove the washers, and your plant should be firmly mounted on the paper.

Plastic Cement Method: If the plastic is not already in some sort of squeeze container, place it in an oil-can squeeze-gun. You may need to mix the plastic with a little thinner so that the plastic will squeeze out readily. Pick the plant up that is to be mounted and squeeze the plastic out on the underside in little stringers along the stem, leaves, and flowers. Next, place the plant on the mounting paper and place washers on as described above. Be sure that you have no washer or weight directly over places where the glue is exposed. If you have unusually large or woody specimens, plastic may also be placed over the top of the stem to act as reinforcement. Do not leave washers on the mounted plant too long, because the washers may stick to delicate parts of the plant, causing them to rip when you try to remove the washers.

Other Methods: If you do not care to glue your plants to the paper, you may use other methods. Many people use tape, either cloth or a firm scotch tape. Cut the tape into narrow strips and place them over the plant in appropriate places until the plant is firmly held on the paper. Your pressed plant must be treated more carefully if you have taped it on because many small parts can be torn off easily.

Labeling

After your plant is mounted, it should be properly labeled. You may only want to include the common and scientific name. Most professional labels are 3 inches by 5 inches in size, and are placed in the lower right corner of the sheet. The label should include the family, genus, species and author(s), the locality including county and state, nature of the habitat, the name of the collector, and the date of the collection. Other information that could be included would be the size of the plant if the entire specimen was not collected, color of the flowers, relative abundance of the plant, elevation of the locality, and associated species.

Final Protection

Pressed and mounted plants are quite delicate and need to be carefully stored. One technique is to place into three-holed plastic sheet protectors. The plants can then be easily placed into a three-holed binder or notebook. This way, your collection can be protected and yet easily viewed by anyone interested. **Note:** make sure your plants are completely dry before placing into plastic sheet protectors, if not they may mold.

GLOSSARY

A

acaulescent flowering stem without leaves; leafless.

achene a dry one-seeded fruit.

acuminate tapering to a sharp point.

acute pointed, but not as much as acuminate.

alternate one side and then the other; not opposite or whorled.

annual grows from seed, flowers, produces seeds, and dies in one year.

anther upper portion of the stamen containing the pollen.

apetalous without petals; sepals may or may not be present.

apex tip.

appendage any supplementary part.

appressed flattened against or close to.

areoles specialized structures of cacti from which the spines and flowers develop.

ascending curving upward in growth.

attenuate slender, tapered point; longer and more gradual than acuminate.

auriculate with ear-like lobes.

awn a bristle, as in the grasses.

axil angle between the leaf and the stem.

axillary borne in the axil.

B

basal occurring at the base.

beak a prolonged, narrow tip of a fruit or a seed.

berry a fleshy fruit.

biennial grows from seed to maturity and dies in two years.

bifid two-parted.

bipinnate twice pinnate.

bipinnatifid two times pinnately divided.

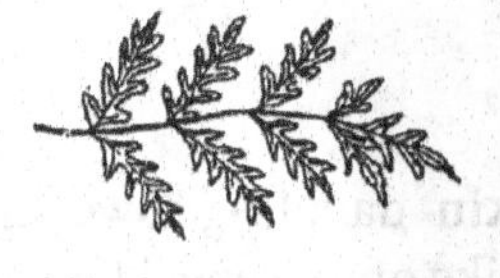

bract leaf-like structure associated with a flower cluster.

bulb underground storage structure composed of stem tip and fleshy leaves.

bur fruit covered with spines.

C

callosity a hardened, thick area.

calyx collective term for the sepals.

campanulate bell shaped.

capillary hair like.

capitate in a dense, head-like cluster.

capsule a dry fruit that opens at maturity.

catkin dangling flower cluster; a spike of unisexual flowers.

caulescent with a leafy stem.

cauline on the stem.

chaparral dense, scrubby vegetation occurring in regions where annual rainfall is between 12 and 25 inches. Most precipitation occurs during winter; summer and fall are extremely dry and hot.

cilia short marginal hairs.

ciliate with a row of fine hairs on the margin.

ciliolate with very fine hairs on the margin.

clasping base of leaf surrounds or partly surrounds the stem.

clawed narrowed at the base.

cleft indented to about the middle, as in leaves.

compound composed of more than one

compound leaf with two or more leaflets.

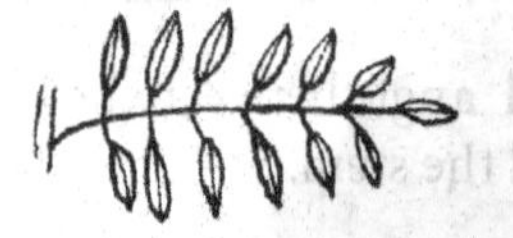

connate united into a single structure.

cordate heart shaped.

corolla collective term for the petals of a flower.

crenate leaf margin with rounded teeth.

crenulate leaf margin with small, rounded teeth.

crisped curled on the margins.

cuneate wedge shaped; often with an indentation at tip.

cyme flat-topped cluster.

D

deciduous falling at maturity or when function is performed.

decumbent flat on ground with slightly raised tip.

decurrent running down the stem.

dentate sharply toothed.

denticulate with fine teeth.

dichotomous branching branching into two equal divisions.

digitate appearing finger like.

dioecious male and female flowers on different plants.

discoid head composite head with only disk flowers.

disk flower type of flower in the *Asteraceae* family.

dissected cut up into smaller segments.

divaricate widely branching or separated.

divided lobed almost to the base.

E

emarginate with a notch at the apex.

entire smooth margined.

epidermis outer layer of tissue.

exfoliating peeling off.

exserted protruding beyond, as stamens protruding beyond the corolla.

F

fascicled clustered in a bunch.

filiform very thin or fine.

fimbriate with a fringed margin.

fleshy thick and somewhat juicy.

floccose with soft woolly tufts.

foliate-leaflet, 3-foliate with 3 leaflets.

fruit matured ovary of a flower; contains the seeds.

funnelform funnel shaped.

G

glabrous without hairs.

glandular with glands; usually sticky or gummy.

glaucous whitened.

globose spherical.

glochids minute barbed bristles of a cactus.

H

habitat place in which the plant lives.

hastate arrow shaped, but with lobes pointed outward.

head dense cluster of sessile flowers.

herb stem not woody.

herbaceous pertaining to an herb.

herbage stem and leaves of the plant.

hirsute with long hairs.

hispid with stiff hairs.

hoary covered with fine white hairs.

hood with a curved structure at the top, like a hood.

host the plant upon which the parasite lives.

hyaline clear or transparent.

I

imbricate shingled or overlapping.

incised sharply indented.

inferior ovary ovary below the petals.

inflorescence clusters of flowers.

involucel a secondary involucre.

involucral bracts bracts surrounding a flower cluster as in the *Asteraceae* family.

involucre bracts beneath a flower cluster, often forming a cup-like structure.

irregular flower flower with bilateral symmetry rather than radial symmetry.

K

keeled with a ridge down the center.

L

laciniate deeply toothed.

lanceolate several times longer than wide.

leaflet one of the leaves of a compound leaf.

legume fruit of the pea plant; usually opens into two halves at maturity.

ligule ray flower of the *Asteraceae* family.

limb spreading portion of a corolla.

linear very narrow; long and thin.

lobe rounded division of a leaf or petal.

M

mealy with a whitish granular substance that can be rubbed off.

merous suffix meaning with parts; for example, 4-merous, with 4 parts.

monocot plants whose seeds have one cotyledon.

mucronate with a short, abrupt tip.

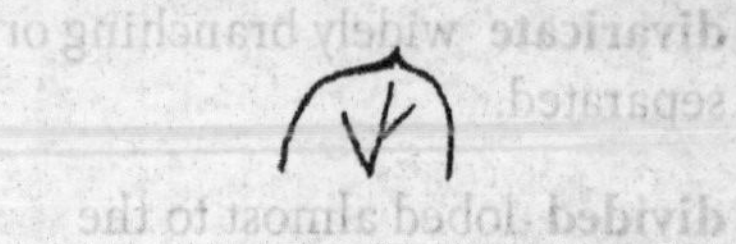

N

nerve a vein or rib.

nettlelike with stinging hairs.

nodding inclined.

node the place on a stem where a leaf or branch arises.

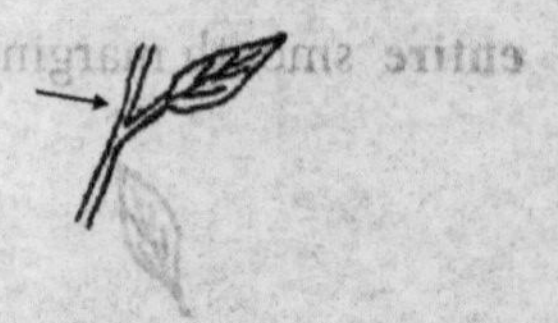

nut a fruit with a hard shell and a single seed.

nutlet a small nut.

O

oblanceolate lanceolate, but with widened portion at the tip.

obovate ovate, with enlarged end near the apex.

obtuse not tapering to a narrow point or tip.

opposite one on each side of a node.

orbicular round or circular.

ovary swollen basal portion of the pistil.

ovate about one and a half times longer than broad. Enlarged end near the petiole (base).

P

paleaceous composed of membranous scales.

palmate finger-like; fanning out from a common point as in leaves.

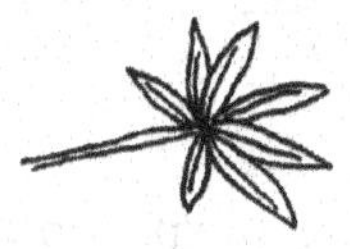

panicle a compound flower cluster.

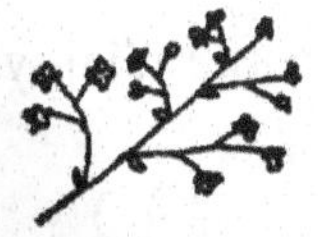

papilionaceous with a flower like that of a sweet pea.

pappus scales or bristles at top of the achene in the *Asteraceae*.

parasite deriving its nourishment from another, with ultimate harm to the host. Many parasitic plants lack chlorophyll.

parted divided to below the middle as in a leaf.

pedicel stem of a flower in a flower cluster.

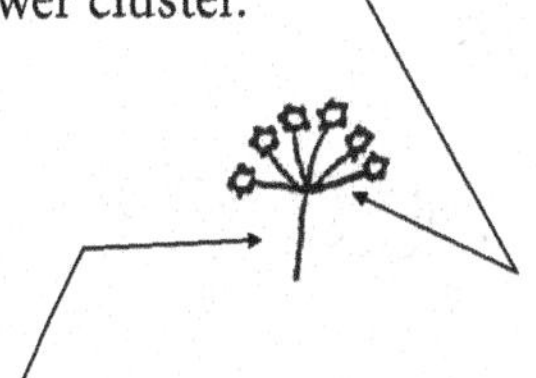

peduncle stem of a flower cluster.

pendulous hanging down.

perennial living for many years.

perfoliate the stem surrounded by the leaf.

perianth floral parts; calyx and corolla, collectively.

petaloid resembling petals.

petiolate with petioles.

petiole stem at the base of a leaf.

pilose with long spreading hairs.

pinnate feather-like; arranged along each side of a common axis.

pinnatifid divided into pinnate segments.

pistil female portion of the flower; will mature into the fruit.

pistillate female.

plumose feathery; with fine hairs on each side.

pod dry fruit; usually elongate and legume-like.

pollen male spores found in the anther of the stamen; a fine powdery substance.

prickle a sharp-pointed projection from the outer layer of the stem or leaf.

procumbent trailing on the ground.

prostrate lying flat on the ground.

puberulent minutely pubescent.

pubescent covered with soft, short hairs.

pulverulent covered with a fine powdery or chalky substance.

R

raceme flower cluster; flowers pedicelled on either side of a main axis.

radiate spreading out in a star-like fashion from a common center; with rays.

ray flower a flower of the *Asteraceae* family.

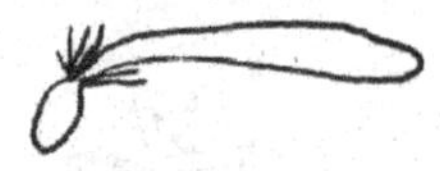

receptacle the portion of a plant which bears the flowers; the base of the flowers or flower.

reflexed turned downward.

regular with radial symmetry.

reniform kidney shaped.

retrorse pointing backward.

revolute rolled inward.

rhizome horizontal underground stem.

rib a ridge or vein.

rosette rose-like cluster at the base; usually referring to leaves.

rotate wheel-shaped.

rugose crinkled or wrinkled texture.

S

sagittate arrow shaped.

salverform with a long, thin, tubular base and expanded upper part.

samara winged fruit.

scabrous rough or sand papery to the touch.

scape leafless flowering stem.

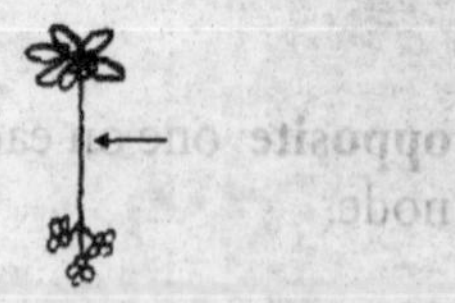

scarious thin, membranous.

scurfy covered with a whitish mealy or granular substance.

segment portion or part, as of a divided leaf.

sepal green, leafy portion of a flower; occurs beneath the petals.

serrate toothed; teeth pointing forward.

sessile without petioles or pedicels.

setose covered with bristles.

sheath base of leaf circling the stem, as in the grasses.

shrub a woody plant; smaller than a tree.

silique a narrow, many-seeded capsule.

simple occurring singly, as a leaf; or without branching, as a stem.

sinuate with a strongly wavy margin.

solitary occurring singly.

sordid dull or brownish.

spatulate (spathulate) somewhat wedge-shaped.

spike inflorescence of sessile flowers.

spine a sharp-pointed structure.

spinescent spine tipped.

spinose having spines.

spreading growing outward from the main stem.

spur slender extension of part of a structure, such as a petal in larkspur.

squarrose with recurving tips.

stamen the male portion of the flower; contains the pollen.

staminate male; with stamens, but no pistils.

stellate star-like; radiating from a central point.

sterile without male or female organs.

stipules small leaf-like structures occasionally found at the base of the petiole.

strigose covered with straight, flat-lying hairs.

style upper narrow neck of the pistil; below the stigma and above the ovary.

subhirsute hairs slightly rigid or stiff.

subsessile with very short petioles or pedicels.

succulent thick and fleshy; juicy.

superior above; superior ovary is when the ovary is placed above the petals and sepals.

symmetrical divisible into two or more equal parts.

T

tendril spiral or cork screw structure arising from the stem; enables climbing.

terete cylindrical.

thorn a sharp-pointed structure; a modified branch.

throat expanded portion of a flower between tube and limb.

tomentose woolly.

tooth a small, marginal lobe; may be rounded or pointed.

trichotomous branching branching into three's.

trifoliate with three leaflets.

tripinnate three time pinnate.

truncate blunt tipped, appearing chopped off.

tube narrow portion at the base of the flower.

turbinate top-shaped.

two-lipped irregular flower with an upper and lower lobe.

U

umbel flower cluster in which pedicelled flowers rise from a single point.

undulate wavy.

unisexual flower flower with stamens or pistil but not both.

V

vein ridge in a leaf which carries the conducting tissue.

venation referring to the arrangement of veins.

villous with many soft hairs.

virgate straight and erect.

viscid sticky.

W

whorl more than two parts radiating out from a point.

winged with a thin, papery extension.

woolly with a mat of soft, tangled hairs.

BIBLIOGRAPHY

Abrams, Leroy, 1959, *Illustrated Flora of the Pacific States*, Stanford University Press, Stanford, 211 pp.

Allen, Robert L., and Fred M. Roberts, Jr., 2013, *Wildflowers of Orange County and the Santa Ana Mountains*, Laguna Wilderness Press, Laguna Beach, 500 pp.

Armstrong, Margaret, 1915, *Field Book of Western Wild Flowers*, C P. Putnam's Sons, New York, 596 pp.

Arnberger, Leslie P., 1952, *Flowers of the Southwest Mountains*, Southwestern Monuments Association, Santa Fe, 112 pp.

Baerg, Harry, 1955, *How to Know the Western Trees*, William C. Brown Company, Dubuque, 170 pp.

Bailey, L. H., 1949, *Manual of Cultivated Plants*, The MacMillan Company, New York, 1,116 pp.

Belzer, Thomas, 1984, *Roadside Plants of Southern California*, Mt. Press Publishing Co., Missoula, 158 pp.

Baldwin, B. G., D. H. Goldman, D. J. Keil, R. Patterson, T. J. Rosatti, and D. H. Wilken, editors. 2012. *The Jepson Manual: Vascular Plants of California*, second edition. University of California Press, Berkeley, 1600 pp.

Balls, Edward K., 1965, *Early Uses of California Plants*, University of California Press, Berkeley, 103 pp.

Benson, Lyman, 1969, *Native Cacti of California*, Stanford University Press, Stanford, 243 pp.

Benson, Lyman, and Robert Darrow, 1945, *A Manual of Southwestern Trees and Shrubs*, University of Arizona, Tucson, 437 pp.

Benson, Lyman, and Robert Darrow, 1954, *The Trees and Shrubs of the Southwestern Deserts*, University of Arizona, Tucson, 437 pp.

California Department of Natural Resources, 1955, *Forests of California*, 147 pp.

Clarke, O. F., D. Svehla, G. Ballmer, and A. Montalvo. 2007. *Flora of the Santa Ana River and Environs with References to World Botany.* Heyday Books, Berkeley, 512 pp.

Collins, Barbara J., 1974, *Key to Trees and Wildflowers of the Mountains of Southern California*, California State University Northridge Foundation, Northridge, CA 91330, 277 pp.

Collins, Barbara J., 1976, *Key to Trees and Shrubs of the Deserts of Southern California*, California State University Northridge Foundation, Northridge, CA 91330, 150 pp.

Collins, Barbara J., 1979, *Key to the Wildflowers of the Deserts of Southern California*, California State University Northridge Foundation, Northridge, CA 91330, 143 pp.

Cornell, Ralph D., 1938, *Conspicuous California Plants*, San Pasqual Press, Pasadena, 192 pp.

Dawson, E. Yale, 1963, *How to Know the Cacti*, William C. Brown Company, Dubuque, 158 pp.

Dawson, E. Yale, 1966, *Cacti of California*, University of California Press, Berkeley, 64 pp.

Dodge, Natt N., 1954, *Flowers of the Southwest Deserts*, Southwestern Monuments Association, Globe, 112 pp.

Fultz, Francis, M., 1927, *The Elfin Forest of California*, The Times-Mirror Press, Los Angeles, 277 pp.

Haskins, Leslie L., 1967, *Wild Flowers of the Pacific Coast*, Binsford and Mort, Portland, OR, 407 pp.

Hickman, J. C., ed., 1993, *The Jepson Manual: Higher Plants of California*, University of California Press, Berkeley, 1400 pp.

Jaeger, Edmund C., 1940, *Desert Wild Flowers*, Stanford University Press, Stanford, 322 pp.

Junak, K. S., et al., 1995, *A Flora of Santa Cruz Island*, Santa Barbara Botanic Garden, Santa Barbara, CA, 397 pp.

Kane, Charles W., 2013, *Southern California Food Plants: Wild Edibles of the Valleys, Foothills, Coast, and Beyond*, Lincoln Town Press, Tucson, 52 pp.

Kirk, Donald R., 1975, *Wild Edible Plants of Western North America*, Naturegraph Publishers, Happ Camp, California, 307 pp.

McAuley, Milt, 1985, *Wildflowers of the Santa Monica Mountains*, Canyon Publishing Company, 8561 Eatough Avenue, Canoga Park, California, 91304, 544 pp.

McMinn, Howard D., 1951, *An Illustrated Manual of California Shrubs*, University of California Press, Berkeley, 409 pp.

Mason, Herbert L., 1957, *A Flora of the Marshes of California*, University of California Press, Berkeley, 878 pp.

Munz, Philip A., 1961, *California Spring Wildflowers*, University of California Press, Berkeley, 122 pp.

Munz, Philip A., 1962, *California Desert Wildflowers*, University of California Press, Berkeley, 122 pp.

Munz, Philip A., 1963, *California Mountain Wildflowers*, University of California Press, Berkeley, 122 pp.

Munz, Philip A., 1964, *Shore Wildflowers of California, Oregon, and Washington*, University of California Press, Berkeley, 122 pp.

Munz, Philip A., 1974, *A Flora of Southern California*, University of California Press, Berkeley, 1,086 pp.

Munz, Philip A., and David A. Keck, 1959, *A California Flora*, University of California Press, Berkeley, 1,681 pp.

Parsons, Mary E., 1955, *The Wild Flowers of California*, California Academy Of Sciences, San Francisco, 423 pp.

Patraw, Pauline M., 1953, *Flowers of the Southwest Mesas*, Southwestern Monuments Association, Globe, 112 pp.

Peattie, D. C., 1953, *A Natural History of Western Trees*, Houghton Mifflin, Boston, 751 pp.

Peterson, P. Victor, 1966, *Native Trees of Southern California*, University of California Press, Berkeley, 136 pp.

Plummer, Fred G., 1911, *Chaparral; Studies in the Dwarf Forests or Elfinwood Of Southern California*, U. S. Forest Service Bulletin 85, 48 pp.

Pohl, R. W., 1954, *How to Know the Grasses*, William C. Brown Company, Dubuque, 192 pp.

Rasmussen, Dean L., 1976, *How to Live Through a Famine*, Horizon Publishers, Bountiful, Utah, 443 pp.

Raven, Peter H., 1966, *Native Shrubs of Southern California*, University of California Press, Berkeley, 132 pp.

Raven, P. H. and Thompson, H. S., 1977, *Flora of the Santa Monica Mountains, California*, University of California, Los Angeles, 189 pp.

Robbins, W. W., Margaret K. Bellue, and Walter S. Ball, 1951, *Weeds of California*, State of California, Sacramento, 547 pp.

Shreve, F., 1951, *Vegetation and Flora of the Sonoran Desert*, Carnegie Institute of Washington, 192 pp.

Sudworth, George, 1908, *Forest Trees of the Pacific Slopes*, U. S. D. A. Forest Service, Washington, D.C., 441 pp.

Ward, Grace B., and Onas M. Ward, *Colorful Desert Wildflowers*, 1978, Best-West Publications, P. O. Box 757, Palm Desert, California, 130 pp.

Parsons, Mary E., 1955, *The Wild Flowers of California*, California Academy Of Sciences, San Francisco, 423 pp.

Patraw, Pauline M., 1953, *Flowers of the Southwest Mesas*, Southwestern Monuments Association, Globe, 112 pp.

Peattie, D. C., 1953, *A Natural History of Western Trees*, Houghton Mifflin, Boston, 751 pp.

Peterson, P. Victor 1966, *Native Trees of Southern California*, University of California Press, Berkeley, 136 pp.

Plummer, Fred G., 1911, *Chaparral: Studies in the Dwarf Forests or Elfinwood Of Southern California*, U.S. Forest Service Bulletin 85, 48 pp.

Pohl, R. W., 1954, *How to Know the Grasses*, William C. Brown Company, D[illegible], 192 pp.

Rasmussen, Dean L., 1976, *How to Live Through a Famine*, Horizon Pub[illegible], Bountiful, Utah, 443 pp.

Raven, Peter H., 1966, *Native Shrubs of Southern California*, [illegible] California Press, Berkeley, 132 pp.

Raven, P. H. and Thompson, H. S., 1977, *Flora of the Sa[illegible] M[illegible]tains, California*, University of California, Los Angeles, 189 pp.

Robbins, W. W., Margaret K. Bellue, and Walter S. Ball, 19[illegible] California, State of California, Sacramento, 547 pp.

Shreve, F., 1951, *Vegetation and Flora of the Sonoran Desert*, [illegible] Institute of Washington, 192 pp.

Sudworth, George, 1908, *Forest Trees of the Pacific Slopes*, U.S.[illegible]A. Forest Service, Washington, D.C., 441 pp.

Ward, Grace B., and Onas M. Ward, *Colorful Desert Wildflowers*, 1978, Best West Publications, P.O. Box 757, Palm Desert, California, 130 pp.

HELPFUL WEBSITES FOR PLANT IDENTIFICATION

CalPhotos: Plants: http://calphotos.berkeley.edu//flora/

Eudicots of Orange County, California: http://darwin.bio.uci.edu/~pjbryant/biodiv/plants/index.htm

Jepson eFlora: http://ucjeps.berkeley.edu/IJM.html

Magnoliids of Orange County: http://nathistoc.bio.uci.edu/plants/Magnoliids.htm

Monocots of Orange County: http://nathistoc.bio.uci.edu/plants/monocots.htm

vPlants—Plant Glossary: http://www.vplants.org/plants/glossary/

Wildflowers and Other Plants of Southern California: http://www.calflora.net/bloomingplants/index.html

Wildflowers of the Santa Monica Mountains National Recreation Area: http://www.smmflowers.org/bloom/bloom.htm

Wildflowers of the Southern California: https://earth.callutheran.edu/Academic_Programs/Departments/Biology/Wildflowers/index.htm

Botanical Gardens And Nature Clubs

Barbara Collins Arboretum at California Lutheran University: https://earth.callutheran.edu/plants/

California Native Plant Society: http://www.cnps.org/

Descanso Gardens: https://www.descansogardens.org/

Eaton Canyon Natural Area: http://www.ecnca.org/

Los Angeles County Arboretum and Botanic Garden: http://www.arboretum.org/

Southern California Botanists: http://www.socalbot.org/index.php

Rancho Santa Ana Botanic Garden: http://www.rsabg.org/

The Huntington Library, Art Collections, and Botanical Gardens: http://www.huntington.org/

The Theodore Payne Foundation: http://theodorepayne.org/

HELPFUL WEBSITES FOR PLANT IDENTIFICATION

CalPhotos: Plants: http://calphotos.berkeley.edu//flora/

Eudicots of Orange County, California: http://darwin.bio.uci.edu/~pjbryant/biodiv/plants/index.htm

Jepson eFlora: http://ucjeps.berkeley.edu/IJM.html

Magnoliids of Orange County: http://nathistoc.bio.uci.edu/plants/Magnoliids.htm

Monocots of Orange County: http://nathistoc.bio.uci.edu/plants/monocots.htm

vPlants—Plant Glossary: http://www.vplants.org/plants/glossary/

Wildflowers and Other Plants of Southern California: http://www.calflora.net/bloomingplants/index.html

Wildflowers of the Santa Monica Mountains National Recreation Area: http://www.smmflowers.org/bloom/bloom.htm

Wildflowers of the Southern California: https://earth.callutheran.edu/Academic_Programs/Departments/Biology/Wildflowers/index.htm

Botanical Gardens And Nature Clubs

Barbara Collins Arboretum at California Lutheran University: https://earth.callutheran.edu/plants/

California Native Plant Society: http://www.cnps.org/

Descanso Gardens: https://www.descansogardens.org/

Eaton Canyon Natural Area: http://www.ecnca.org/

Los Angeles County Arboretum and Botanic Garden: http://www.arboretum.org/

Southern California Botanists: http://www.socalbot.org/index.php

Rancho Santa Ana Botanic Garden: http://www.rsabg.org/

The Huntington Library, Art Collections, and Botanical Gardens: http://www.huntington.org/

The Theodore Payne Foundation: http://theodorepayne.org/

PLANT ID CHECKLIST

- ☐ alfalfa
- ☐ alkali heath
- ☐ alkali-mallow
- ☐ American brooklime
- ☐ American sea rocket
- ☐ American vetch
- ☐ annual bur-sage
- ☐ annual malacothrix
- ☐ antisell milkvetch
- ☐ arrow-weed
- ☐ arroyo lupine
- ☐ arroyo willow
- ☐ Australian cotula
- ☐ Australian saltbush
- ☐ baby blue-eyes
- ☐ bajada lupine
- ☐ beach bur-sage
- ☐ beach evening-primrose
- ☐ beach morning-glory
- ☐ beach saltbush
- ☐ beavertail
- ☐ Bermuda buttercup
- ☐ biennial wormword
- ☐ big leaf mistletoe
- ☐ big saltbush
- ☐ bigberry manzanita
- ☐ Bigelow's tickseed
- ☐ big-leaf maple
- ☐ bigpod ceanothus
- ☐ bindweed
- ☐ birch-leaf mountain-mahogany
- ☐ bird's-foot trefoil
- ☐ black cottonwood
- ☐ black mustard
- ☐ black sage
- ☐ bladderpod
- ☐ Blochman's dudleya
- ☐ blow-wives
- ☐ blue dicks
- ☐ blue elderberry
- ☐ bluehead gilia
- ☐ bluff lettuce
- ☐ Boccone's sand-spurrey
- ☐ branching phacelia
- ☐ brass-buttons
- ☐ Brewer's ragwort
- ☐ bristly bird's beak
- ☐ bristly ox-tongue
- ☐ broad-leafed lupine
- ☐ broad-leaved cattail
- ☐ brown dogwood
- ☐ buckbrush
- ☐ buffalo gourd or calabazilla
- ☐ bull mallow
- ☐ bull thistle
- ☐ burhead
- ☐ bush poppy
- ☐ bushrue
- ☐ butterfly mariposa lily
- ☐ California ash
- ☐ California aster
- ☐ California barberry
- ☐ California bay
- ☐ California blackberry
- ☐ California brickellbush
- ☐ California brittlebush
- ☐ California buckeye
- ☐ California buckwheat
- ☐ California burclover
- ☐ California buttercup
- ☐ California centaury
- ☐ California chicory
- ☐ California coffee berry
- ☐ California cottonrose
- ☐ California croton
- ☐ California evening-primrose
- ☐ California everlasting
- ☐ California false-indigo
- ☐ California figwort
- ☐ California fuchsia
- ☐ California goldenrod
- ☐ California goldfields
- ☐ California goosefoot
- ☐ California hedge nettle
- ☐ California hummingbird sage
- ☐ California loosestrife
- ☐ California milkweed
- ☐ California mustard
- ☐ California peony
- ☐ California plantain
- ☐ California poppy
- ☐ California ragwort
- ☐ California rockcress
- ☐ California rose
- ☐ California sagebrush
- ☐ California saxifrage
- ☐ California sun cup
- ☐ California thistle
- ☐ canyon clarkia
- ☐ canyon dodder
- ☐ canyon live oak
- ☐ canyon pea
- ☐ canyon silktassel
- ☐ canyon sunflower
- ☐ castor bean
- ☐ Catalina mariposa lily
- ☐ caterpillar phacelia
- ☐ celery
- ☐ chalk dudleya
- ☐ chamise
- ☐ chaparral clematis
- ☐ chaparral currant
- ☐ chaparral dodder
- ☐ chaparral gilia
- ☐ chaparral mallow
- ☐ chaparral pea
- ☐ chaparral prickly-pear
- ☐ chaparral star lily
- ☐ chaparral yucca
- ☐ checkerbloom
- ☐ cheeseweed
- ☐ chia
- ☐ chicory
- ☐ chicoryleaf wire-lettuce
- ☐ Chinese-houses
- ☐ chocolate lily

- ☐ Cleveland's cryptantha
- ☐ Cleveland's tobacco
- ☐ coast boykinia
- ☐ coast cholla
- ☐ coast goosefoot
- ☐ coast Indian paintbrush
- ☐ coast live oak
- ☐ coast prickly-pear
- ☐ coast silktassel
- ☐ coast woolly-heads
- ☐ coastal goldenbush
- ☐ coastal lotus
- ☐ coastal wild buckwheat
- ☐ cobwebby thistle
- ☐ cocklebur
- ☐ collar lupine
- ☐ common beggar-ticks
- ☐ common chickweed
- ☐ common dandelion
- ☐ common eucrypta
- ☐ common fiddleneck
- ☐ common goldenstar
- ☐ common groundsel
- ☐ common gumplant
- ☐ common hillside daisy
- ☐ common knotweed
- ☐ common lomatium
- ☐ common madia
- ☐ common muilla
- ☐ common phacelia
- ☐ common plantain
- ☐ common pussypaws
- ☐ common sow thistle
- ☐ common sunflower
- ☐ Conejo buckwheat
- ☐ Conejo dudleya
- ☐ Coulter's lupine
- ☐ Coulter's matilija poppy
- ☐ coyote brush
- ☐ cream cups
- ☐ creeping snowberry
- ☐ creeping wood-sorrel
- ☐ crete weed
- ☐ crystalline iceplant
- ☐ curly dock
- ☐ deer brush
- ☐ deerpea vetch
- ☐ deerweed
- ☐ delicate clarkia
- ☐ Del Mar manzanita
- ☐ deltoid balsamroot
- ☐ dense false gilia
- ☐ diamond clarkia
- ☐ dotted smartweed
- ☐ Douglas' microseris
- ☐ Douglas' stitchwort
- ☐ Douglas' iris
- ☐ downy monkeyflower
- ☐ dune lupine
- ☐ dwarf athysanus
- ☐ dwarf coastweed
- ☐ dwarf nettle
- ☐ early onion
- ☐ Eastwood manzanita
- ☐ elegant clarkia
- ☐ elegant silverpuffs
- ☐ Emory's rock daisy
- ☐ Engelmann oak
- ☐ English plantain
- ☐ European sea rocket
- ☐ everlasting neststraw
- ☐ fascicled tarweed
- ☐ fat-hen
- ☐ feltleaf/white everlasting
- ☐ felt-leaved monardella
- ☐ fennel
- ☐ field mint
- ☐ field mustard
- ☐ field willowherb
- ☐ fiesta flower
- ☐ fire poppy
- ☐ five-hook bassia
- ☐ flax-leaved horseweed
- ☐ fleshy pincushion
- ☐ flix weed
- ☐ foothill clover
- ☐ foothill penstemon
- ☐ fragrant everlasting
- ☐ fragrant pitcher sage
- ☐ freeway iceplant
- ☐ Fremont cottonwood
- ☐ French broom
- ☐ fringed spineflower
- ☐ fringe-pod
- ☐ fuchsia-flowered gooseberry
- ☐ gallant soldier
- ☐ Gambel milkvetch
- ☐ garland daisy
- ☐ giant blazing star
- ☐ giant tickseed
- ☐ golden crownbeard
- ☐ golden currant
- ☐ golden eardrops
- ☐ golden-fleece
- ☐ golden-yarrow
- ☐ Goodding's black willow
- ☐ goose grass
- ☐ grape soda lupine
- ☐ greenbark ceanothus
- ☐ greenstem filaree
- ☐ ground pink
- ☐ gumweed
- ☐ hairy ceanothus
- ☐ hairy evening-primrose
- ☐ hairy matilija poppy
- ☐ hairy sand-spurrey
- ☐ hairy vetch
- ☐ heart-leaved bush penstemon
- ☐ heart-podded hoary cress
- ☐ hedge mustard
- ☐ Heermann's lotus
- ☐ hillside woodland star
- ☐ hoary fuchsia
- ☐ hoary nettle
- ☐ hoaryleaf ceanothus
- ☐ hollyleaf navarretia
- ☐ hollyleaf redberry
- ☐ holly-leaved cherry
- ☐ hooked navarretia
- ☐ Hooker's evening-primrose
- ☐ horehound
- ☐ horseweed
- ☐ Humboldt lily
- ☐ Indian tobacco
- ☐ interior goldenbush
- ☐ interior live oak
- ☐ Italian thistle
- ☐ Jimson weed "sacred datura"
- ☐ Johnny-jump-up
- ☐ lace-pod
- ☐ lamb's quarters

- ☐ lance-leaved dudleya
- ☐ large-flowered lotus
- ☐ large-flowered phacelia
- ☐ largefruit amaranth
- ☐ laurel sumac
- ☐ leafy fleabane
- ☐ lemonade berry
- ☐ lesser paintbrush
- ☐ Lindley's silverpuffs
- ☐ London rocket
- ☐ long-beaked storksbill
- ☐ long-stem wild buckwheat
- ☐ lowland cudweed
- ☐ Lyon's pentachaeta
- ☐ many-stemmed dudleya
- ☐ marcescent dudleya
- ☐ marsh jaumea
- ☐ mayweed
- ☐ meadow-rue
- ☐ mesa brodiaea
- ☐ Mexican pink
- ☐ Mexican tea
- ☐ milk maids
- ☐ milk thistle
- ☐ miner's lettuce
- ☐ miniature lupine
- ☐ mission prickly-pear
- ☐ mock heather
- ☐ mountain dandelion
- ☐ mugwort
- ☐ mule fat
- ☐ musk monkeyflower
- ☐ mustang mint
- ☐ mustard evening-primrose
- ☐ narrow-leaf milkweed
- ☐ narrow-leaved cattail
- ☐ narrow-leaved willow
- ☐ narrowly leaved bedstraw
- ☐ narrow-toothed pectocarya
- ☐ nettleleaf goosefoot
- ☐ New Zealand spinach
- ☐ northern pectocarya
- ☐ Nuttall's scrub oak
- ☐ oak mistletoe
- ☐ oceanspray
- ☐ oriental mustard
- ☐ Pacific pickleweed
- ☐ Pacific silverwood
- ☐ Pacific willow
- ☐ Parry's larkspur
- ☐ Parry's phacelia
- ☐ peak rush-rose
- ☐ pearly everlasting
- ☐ peninsular onion
- ☐ pennyroyal
- ☐ Persian knotweed
- ☐ petty spurge
- ☐ pineapple weed
- ☐ pine-bush
- ☐ pink everlasting
- ☐ pink sand-verbena
- ☐ pinpoint clover
- ☐ Plummer's mariposa lily
- ☐ poison hemlock
- ☐ prickly lettuce
- ☐ prickly phlox
- ☐ prickly sow thistle
- ☐ prince's plume
- ☐ procumbent pigweed
- ☐ prostrate spineflower
- ☐ punchbowl godetia
- ☐ puncture vine
- ☐ purple clarkia
- ☐ purple nightshade
- ☐ purple owl's clover
- ☐ purple sage
- ☐ purple sanicle
- ☐ purslane
- ☐ pygmy-weed
- ☐ q-tips
- ☐ radish
- ☐ Ramona clarkia
- ☐ rattlesnake weed
- ☐ red clover
- ☐ red flax
- ☐ red maids
- ☐ red sand-verbena
- ☐ red shank
- ☐ red willow
- ☐ redroot pigweed
- ☐ red-skinned onion
- ☐ redstem filaree
- ☐ robust vervain
- ☐ rock malacothrix
- ☐ rod wire-lettuce
- ☐ rose snapdragon
- ☐ round-leaved boykinia
- ☐ royal goldfields
- ☐ Russian thistle
- ☐ rusty popcornflower
- ☐ sacapellote
- ☐ salt dodder
- ☐ salt marsh bird's-beak
- ☐ saltmarsh fleabane
- ☐ saltmarsh sand-spurrey
- ☐ San Diego or climbing bedstraw
- ☐ San Diego viguiera
- ☐ sand/sea lettuce
- ☐ Santa Monica dudleya
- ☐ Santa Susana tarplant
- ☐ saw-toothed goldenbush
- ☐ scalebroom
- ☐ scarlet bugler
- ☐ scarlet monkeyflower
- ☐ scarlet pimpernel
- ☐ scarlet/cardinal larkspur
- ☐ Scotch broom
- ☐ scrub oak
- ☐ sea fig
- ☐ seacliff wild buckwheat
- ☐ seaside heliotrope
- ☐ seep-spring monkeyflower
- ☐ sessileflower goldenaster
- ☐ shepherd's purse
- ☐ shining peppergrass
- ☐ shiny lomatium
- ☐ shooting star
- ☐ shortpod mustard
- ☐ showy penstemon
- ☐ shredding evening-primrose
- ☐ silver bush lupine
- ☐ silver lotus
- ☐ skullcap
- ☐ skunk bush
- ☐ slender sunflower
- ☐ slender-leaved iceplant
- ☐ slender-spire orchid
- ☐ small evening-primrose
- ☐ small-flowered meconella
- ☐ small-flowered nightshade
- ☐ smallseed sandmat

- ☐ smooth cat's-ear
- ☐ snakeroot
- ☐ soap plant
- ☐ sourclover
- ☐ southern CA black walnut
- ☐ southern California milkvetch
- ☐ southern honeysuckle
- ☐ southern mountain misery
- ☐ southern tauschia
- ☐ Spanish broom
- ☐ Spanish clover
- ☐ spearmint
- ☐ speckled clarkia
- ☐ spiny cocklebur
- ☐ spiny redberry
- ☐ splendid mariposa lily
- ☐ spotted spurge
- ☐ spreading larkspur
- ☐ spring vetch
- ☐ stickleaf
- ☐ stickwort
- ☐ sticky cinquefoil
- ☐ sticky mouse-ear chickweed
- ☐ sticky sand-spurrey
- ☐ sticky/bush monkeyflower
- ☐ stinging lupine
- ☐ stream orchid
- ☐ strigose lotus
- ☐ strigose sun cup
- ☐ sugar bush
- ☐ sweet alyssum
- ☐ sweet-cicely
- ☐ tansy phacelia
- ☐ tarragon
- ☐ telegraph weed
- ☐ thickbracted goldenbush
- ☐ thick-leaved yerba santa
- ☐ thin-leaf malacothrix
- ☐ thistle sage
- ☐ threadleaf ragwort
- ☐ thyme-leafed spurge
- ☐ tidy-tips
- ☐ tocalote
- ☐ tomcat clover
- ☐ toyon
- ☐ tree tobacco
- ☐ tufted poppy
- ☐ tumble mustard
- ☐ tumbleweed
- ☐ tumbling orach
- ☐ turkey mullein "doveweed"
- ☐ Turkish rugging
- ☐ twining snapdragon
- ☐ two-tone everlasting
- ☐ valley cholla
- ☐ valley oak
- ☐ valley popcornflower
- ☐ variable linanthus
- ☐ vinegar weed
- ☐ violet snapdragon
- ☐ warrior's plume
- ☐ water cress
- ☐ water smartweed
- ☐ water speedwell
- ☐ wax myrtle
- ☐ weakleaf bur ragweed
- ☐ wedge-leaf horkelia
- ☐ Weed's mariposa lily
- ☐ western blue-eyed-grass
- ☐ western columbine
- ☐ western goldenrod
- ☐ western poison oak
- ☐ western ragweed
- ☐ western sycamore
- ☐ western tansy mustard
- ☐ western vervain
- ☐ western virgin's bower
- ☐ western wallflower
- ☐ whispering bells
- ☐ white alder
- ☐ white clover
- ☐ white eardrops
- ☐ white globe lily
- ☐ white hedge nettle
- ☐ white layia
- ☐ white nightshade
- ☐ white pincushion
- ☐ white pitcher sage
- ☐ white rabbit-tobacco
- ☐ white sage
- ☐ white snapdragon
- ☐ white sweetclover
- ☐ white-leaf monardella
- ☐ wide-throated YL monkeyflower
- ☐ wild Canterbury bells
- ☐ wild celery
- ☐ wild cucumber or chilicothe
- ☐ wild licorice
- ☐ wild morning-glory
- ☐ wild-rhubarb
- ☐ willow dock
- ☐ willowherb
- ☐ windmill pink
- ☐ winter vetch
- ☐ wishbone bush
- ☐ woodland threadstem
- ☐ woolly blue curls
- ☐ woolly lomatium
- ☐ woolly paintbrush
- ☐ woolly seablite
- ☐ yarrow
- ☐ yellow bush lupine
- ☐ yellow mariposa lily
- ☐ yellow penstemon
- ☐ yellow pincushion
- ☐ yellow star-thistle
- ☐ yellow water primrose
- ☐ yellowstem bush-mallow
- ☐ yerba buena
- ☐ yerba mansa

INDEX TO SPECIES BY PLANT FAMILIES

A

B

C

E

F

M

N

O

P

R

S

T

U

V

Z

INDEX TO SCIENTIFIC NAMES

D

M

N

O

P

Q

R

S

T

U

V

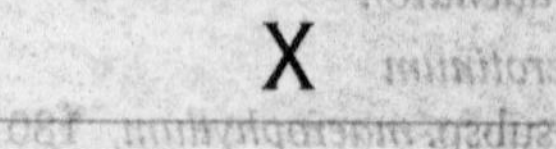

X

Z

INDEX TO COMMON NAMES

A

B

C

D

E

F

G

H

I

J

K

L

Q

R

S

T

V